Gradient Smoothing Methods
with Programming

Applications to Fluids and Landslides

Frontier Research in Computation and Mechanics of Materials and Biology

ISSN: 2315-4713

Series Editors: Shaofan Li *(University of California, Berkeley, USA)*
Wing Kam Liu *(Northwestern University, USA)*
Xanthippi Markenscoff *(University of California, San Diego, USA)*

Frontier Research in Computation and Mechanics of Materials and Biology – Vol. 5

Gradient Smoothing Methods with Programming

Applications to Fluids and Landslides

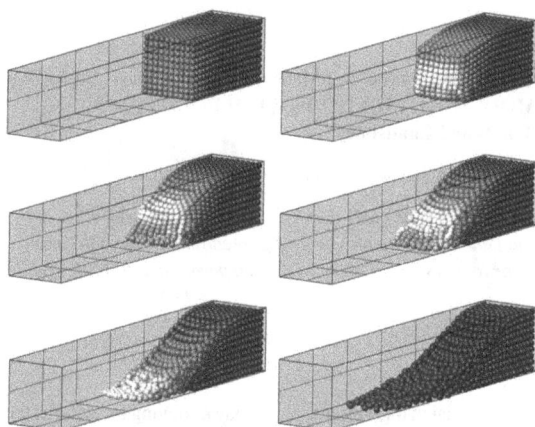

G R Liu
University of Cincinnati, USA

Zirui Mao
Pacific Northwest National Laboratory, USA

World Scientific

NEW JERSEY · LONDON · SINGAPORE · BEIJING · SHANGHAI · HONG KONG · TAIPEI · CHENNAI · TOKYO

Published by

World Scientific Publishing Co. Pte. Ltd.

5 Toh Tuck Link, Singapore 596224

USA office: 27 Warren Street, Suite 401-402, Hackensack, NJ 07601

UK office: 57 Shelton Street, Covent Garden, London WC2H 9HE

Library of Congress Control Number: 2023040958

British Library Cataloguing-in-Publication Data
A catalogue record for this book is available from the British Library.

Frontier Research in Computation and Mechanics of Materials and Biology — Vol. 5
GRADIENT SMOOTHING METHODS WITH PROGRAMMING
Applications to Fluids and Landslides

ISBN 978-981-128-000-9 (hardcover)
ISBN 978-981-128-001-6 (ebook for institutions)
ISBN 978-981-128-002-3 (ebook for individuals)

For any available supplementary material, please visit
https://www.worldscientific.com/worldscibooks/10.1142/13508#t=suppl

Desk Editors: Balasubramanian Shanmugam/Steven Patt

Typeset by Stallion Press
Email: enquiries@stallionpress.com

About the Authors

Gui-Rong Liu received his Ph.D. from Tohoku University, Japan, in 1991. He was a Postdoctoral Fellow at Northwestern University, USA, from 1991–1993. He was a Professor at the National University of Singapore until 2010. He is currently a Professor at the Department of Aerospace Engineering and Engineering Mechanics, University of Cincinnati, USA. He was the Founder of the Association for Computational Mechanics (Singapore) (SACM) and served as the President of SACM until 2010. He served as the President of the Asia-Pacific Association for Computational Mechanics (APACM) (2010–2013) and an Executive Council Member of the International Association for Computational Mechanics (IACM) (2005–2010; 2020–2026). He authored a large number of journal papers and books including two best-sellers: *Mesh Free Method: Moving Beyond the Finite Element Method* and *Smoothed Particle Hydrodynamics: A Meshfree Particle Methods*. He is the Editor-in-Chief of the *International Journal of Computational Methods* and served as an Associate Editor for *IPSE* and *MANO*. He is the recipient of numerous awards, including the Singapore Defence Technology Prize, NUS Outstanding University Researcher Award, NUS Best Teacher Award, APACM Computational Mechanics Award, JSME Computational Mechanics Award, ASME Ted Belytschko Applied Mechanics Award, Zienkiewicz Medal from APACM, the AJCM Computational Mechanics Award, and the

Humboldt Research Award. He has been listed as one among the world's top 1% most influential scientists (Highly Cited Researchers) by Thomson Reuters for a number of years.

Zirui Mao received his Ph.D. from the University of Cincinnati, USA, in 2019 under the supervision of Prof. GR Liu. He was a Postdoctoral Researcher at Texas A&M University, USA, from 2019–2021. He is currently a Research Associate at the Pacific Northwest National Laboratory, USA. His research focuses on computational mechanics, wherein he employs the knowledge of mechanics, mathematics, and emerging machine learning algorithms to simulate complex engineering systems via High Performance Computation (HPC). His research experience involves meshfree particle methods and grid-based numerical methods, the phase field method, Fast Fourier Transformation, and machine learning algorithms with applications in geotechnical engineering, nuclear engineering, material processing engineering, and ocean engineering.

Contents

Chapter 1

Introduction

Contents

This chapter provides a brief background overview on the existing numerical methods, with a focus on meshfree methods [1] that are directly relevant to this book. Key ingredients of numerical modeling and a comparison of numerical methods will be discussed, leading to motivations behind the Gradient Smoothing Methods (GSMs).

The chapter is structured as follows:

- Section 1.1 introduces the *basic concepts* of numerical simulation and its role in scientific research.
- Section 1.2 presents the *key ingredients* and procedures of numerical simulation, using the Finite Difference Method (FDM) as an example.
- Section 1.3 explains how the partial differential equations describing the physical problems are converted to routinely solvable algebraic equations.
- Section 1.4 compares some of the existing major numerical methods, including Lagrangian/Eulerian methods and grid-based/meshfree particle methods.
- Section 1.5 presents the *progress* of meshfree methods, with a focus on the Smoothed Particle Hydrodynamics (SPH). The advantages and limitations of the SPH method are also discussed.
- Section 1.6 gives the *motivations* behind the development of the GSMs.
- Section 1.7 defines the *notations* used throughout the book.

Why this book. This book presents relatively new GSMs to meet the challenges in solving mechanics problems related to fluids and fluid-like solids in both natural events and engineering systems. While the well-known Finite Difference Method (FDM) is the oldest and is still quite commonly used for fluids, it has limitations when the geometry of the problem domain is complicated. Similarly, the Finite Element Method (FEM) is widely used for solids and structures [2], but struggles to handle extremely large flow-like deformation, due to severe element mesh distortion. To overcome these limitations, the GSMs are presented as candidate numerical methods.

The GSMs will be discussed in detail in the following chapters:

- Chapter 2 presents the basic concepts and theories of the gradient smoothing technique, based on which the GSMs are proposed.

- Chapter 3 details the key ingredients of GSM and its effectiveness in solving fluid flow problems with complex geometries.
- Chapter 4 presents applications of GSM to compressible flows, while Chapter 5 presents applications of GSM to incompressible flows.
- Chapter 6 formulates a Lagrangian-GSM (L-GSM) for problems involving fluid-like solids with material strength and large deformation.
- Chapter 7 applies L-GSM to hydrodynamics, while Chapter 8 applies L-GSM to geotechnical engineering.
- Finally, Chapter 9 provides in-house MATLAB codes of both GSM and L-GSM, along with detailed descriptions.

1.1 On numerical simulation

In the past century, significant progress has been made in developing theoretical and experimental methods to solve problems in natural science and engineering. These approaches often need careful assumptions, simplifications, and the creation of scaled-down physical models. The process is also time-consuming and costly, and experimental methods can sometimes even be dangerous.

Rapid development of computer techniques in recent decades has led to the increasing use of computer-based simulations to solve complex problems in engineering and the sciences. Unlike traditional approaches, modern numerical methods allow for the detailed examination of real-world problems without relying on too many assumptions. These methods cast physical problems into systems of equations, which can be solved effectively and routinely on a computer, enabling the visualization of critical phenomena and providing valuable solutions to these problems.

Numerical simulations have demonstrated excellent predictive power, allowing researchers to create vivid simulations of complex physical events, thereby providing a deeper understanding of critical phenomena. This approach provides an alternative tool for scientific investigations, enabling researchers to achieve results that are not achievable otherwise. Furthermore, numerical simulations with carefully designed computational methods play an essential role in providing verification of theories, offering insights into experimental results, and even leading to the discovery of new phenomena. Together with theories and experiments, numerical

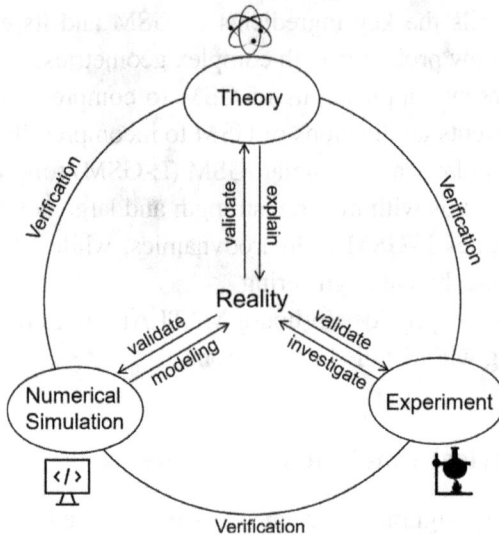

Figure 1.1. Three major pillars in scientific studies: theories, numerical simulations, and experiments.

simulations have become one of the three major pillars in science and technology for solving real-life problems, as shown in Figure 1.1.

1.2　Computational methods

In natural and engineered systems, we encounter various physical phenomena, such as fluid flows (air, water, *etc.*) and solid flows (transport of granular materials, landslides, etc.), that can be analyzed through mechanics. Computational methods simulate these mechanics problems in a virtual environment. It involves the integration of mechanics, mathematics, engineering applications, computer science, and emerging AI algorithms [84], as illustrated in Figure 1.2.

To describe a mechanics problem, a mathematical model must be derived following proper principles and theories. This step involves understanding the mechanics of solids and fluids, as well as the mathematical representations of these models. The mathematical models of solids and fluids are usually in the form of nonlinear dynamical partial differential equations (PDEs) defined in the problem domain, with proper boundary and initial conditions. Most of the PDEs cannot be solved analytically for

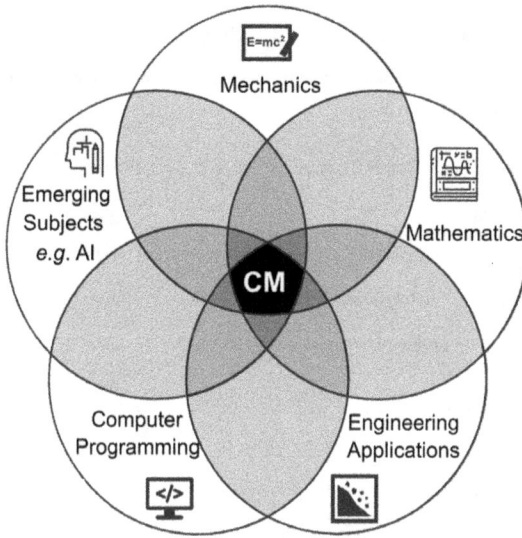

Figure 1.2. Central roles of the computational methods in various disciplines.

an exact solution. Computational methods are thus needed for numerical solutions. This involves discretizing the PDEs and then solving the discretized equation system numerically for a stable solution that converges to the exact solution.

With a proper numerical scheme, effective coding and programming are required for implementation. The numerical solution can then be obtained and visualized for analysis and evaluation via a computer. The entire computational procedure can be represented by Figure 1.3.

In numerical simulations, the problem domain is commonly discretized into small elements, cell, grids, particles, or a form of mesh. Take the widely used Finite Difference Method (FDM) as an example. A typical procedure of numerical simulation can be indicated using the simple and widely studied lid-driven cavity problem.

Physical problem description: The lid-driven cavity problem is illustrated in Figure 1.4(a). It involves circulating flow in a square region created by moving the lid on the top surface, with other three surfaces treated as no-slip solid boundaries. The task is to find the unknown velocity field of the fluid in the entire square domain.

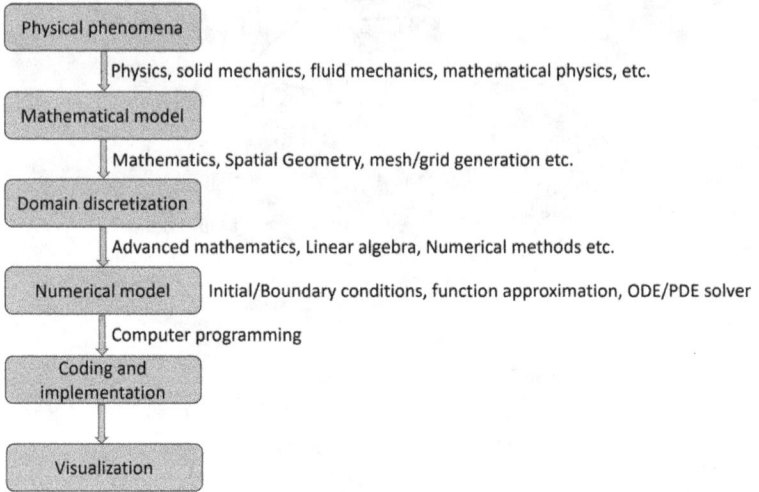

Figure 1.3.   Typical procedure for conducting computer modeling.

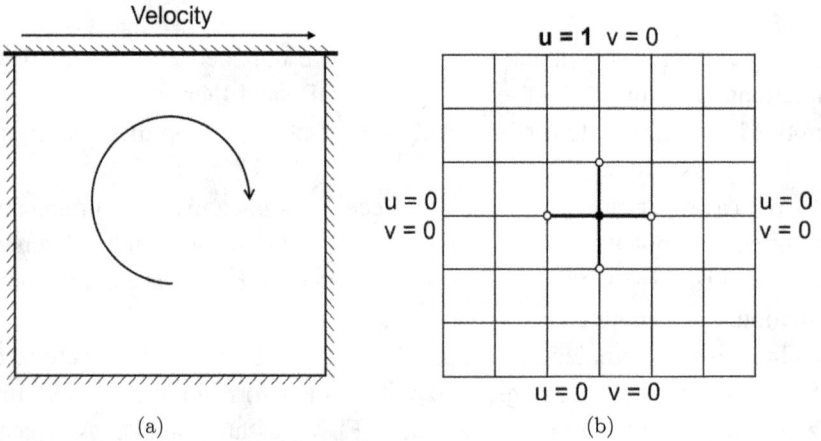

Figure 1.4.   (a) Setting of the lid-driven cavity problem defined in a square domain. (b) Domain discretization with uniform grid with regular cells and uniformly distributed nodes for a finite difference model.

Mathematical model: Based on the knowledge of fluid dynamics, the cavity flow problem is mathematically governed by the well-known Navier–Stokes equations, which are a typical set of PDEs. The PDEs are often accompanied by "equations of state" to define the behavior of a fluid. If the fluid is assumed incompressible and inviscid, the mathematical model has the following simplified form of PDEs [3]:

$$\begin{cases} \omega_t + u\omega_x + v\omega_y = \dfrac{1}{Re}(\omega_{xx} + \omega_{yy}) \\ \psi_{xx} + \psi_{yy} = -\omega \end{cases} \tag{1.1}$$

where u and v are the unknown field variables of velocities in horizontal and vertical directions, respectively. Re is the Reynolds number, which gives a general feature of the flow. The subscripts, t, x, and y, stand for the partial derivatives with respect to time t, x, and y, respectively. ω and ψ are the vorticity and stream function of the flow, respectively, and are given as

$$\begin{cases} \omega = v_x - u_y \\ u = \psi_y; v = -\psi_x \end{cases} \tag{1.2}$$

Domain discretization: To solve the set of PDEs in Eqs. (1.1) and (1.2) numerically, the problem domain is discretized with a grid that consists of cells and nodes, as shown in Figure 1.4(b). The continuous form of PDEs governing the field variables is then approximated with the values of the nodes. The variation of the field variables in the vicinity of a node is approximated with simple functions such as polynomials. Their derivatives can be approximated using a central finite difference (FD) scheme based on the standard Taylor series expansion:

$$\begin{cases} F_x = \dfrac{\partial F}{\partial x} = \dfrac{F_{i+1,j} - F_{i-1,j}}{2\Delta x} \\ F_y = \dfrac{\partial F}{\partial y} = \dfrac{F_{i,j+1} - F_{i,j-1}}{2\Delta y} \end{cases} \tag{1.3}$$

$$\begin{cases} F_{xx} = \dfrac{\partial^2 F}{\partial x^2} = \dfrac{F_{i-1,j} - 2F_{i,j} + F_{i+1,j}}{\Delta x^2} \\ F_{yy} = \dfrac{\partial^2 F}{\partial y^2} = \dfrac{F_{i,j-1} - 2F_{i,j} + F_{i,j+1}}{\Delta y^2} \end{cases} \tag{1.4}$$

Numerical model: Plugging the FD approximation into the governing PDEs (1.1) and (1.2) yields the discretized form of governing equations. This set of equations becomes algebraic with no more differentials with respect to the coordinates, and the unknown field variables are now only at the nodes, rather than continuous functions everywhere in the problem domain.

To solve the set of discrete algebraic equations, the field variables at the nodes on the boundary of the problem domain should be imposed, meaning that we are looking for a solution satisfying the *boundary conditions* (BCs). Boundary conditions of a numerical model often need special care for numerical stability reasons, and the BCs must be satisfied during the whole process. Moreover, it also requires proper handling of the *initial condition* (IC) for transient problems where variables change with time. Initial values of field variables should be assigned to all the nodes at the first time step when the evolution starts.

Programming: Once the numerical model is developed adequately, one needs to have it implemented in codes for computation. Field variables over the whole domain in the entire dynamic process are computed.

Visualization and analysis: Finally, numerical results are visualized (as shown in Figure 1.5) with the help of some visualization software, known as postprocessors.

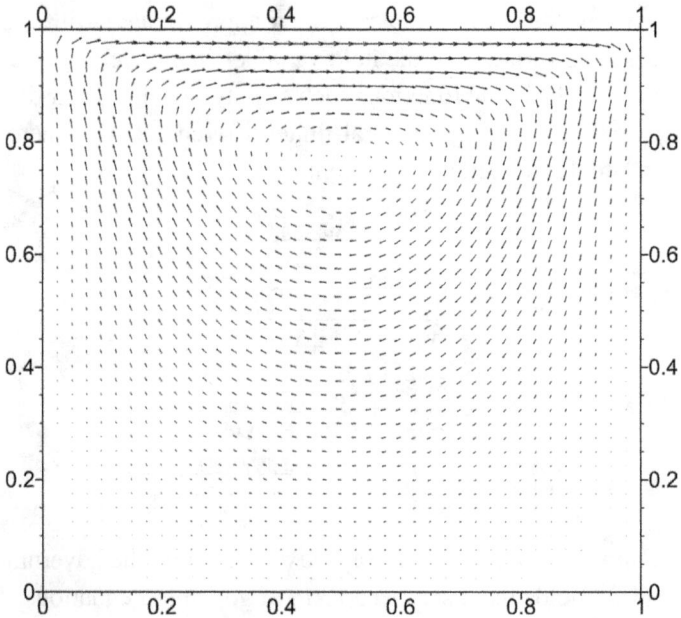

Figure 1.5. Visualization of numerical result obtained using our FDM code for the lid-driven cavity problem. Velocity field is plotted in arrows with magnitudes (length of arrows) and directions for $Re = 100$.

1.3 From physics to mathematics

1.3.1 *Well-posed physical problems*

In this book, we focus on well-posed problems, which refer to physical problems governed by a set of PDEs with boundary conditions that have a unique solution. It is the authors' opinion that if the physic problem does not have a solution, there is nothing a numerical method can do, except providing a false solution. For ill-posed problems, specific techniques are required for meaningful solutions, which are not discussed in this text. Interested readers may refer to [4].

1.3.2 *Differential to algebra*

The key objective of numerical methods is to transform a given partial *differential* equation (PDE) with appropriate boundary conditions (BCs) into a system of *algebraic* equations (SAEs). This allows one to obtain the solution using standard algebraic techniques. Various techniques can be used to achieve this, including transformation, series expansion, and domain discretization. This book focuses exclusively on discretization methods, because of its wide application to practical problems in nature, sciences, and engineering.

For any numerical method, the most critical step is to convert the differentiation of an unknown field variable to a simple summation of the field variable at its surrounding points. This step inevitably introduces errors, as higher-order terms must be truncated when PDEs are converted to SAEs.

An effective numerical method must ensure that the resulting SAEs accurately represent the PDEs, have no numerical artifacts, have a unique solution to a well-posed physical problem, and converge to the exact solution of the PDE with BCs, when the mesh/grids are refined. In summary, the SAEs should exhibit good convergence behaviors with respect to the original PDE.

1.3.3 *Methods and types of PDEs*

PDEs are categorized based on the types of physics problems they represent, which determine their spatial characteristics. For solid mechanics problems, elliptic PDEs are typically used, while fluid dynamics problems are governed by hyperbolic PDEs such as the Navier–Stokes equations. Numerical methods are in general different for different types of PDEs.

For elliptic PDEs, the widely used Finite Element Method (FEM) [2] and more advanced Smoothed FEM (S-FEM) [5] allow for the use of unstructured meshes with elements and nodes for arbitrarily complex geometries. These methods typically use a Lagrangian frame, meaning that the nodes are moving. These methods suppress numerical artifacts and achieve efficacy through the proper use of weak forms or weakened weak (W2) forms [6]. FEM reduces the order of differentials for field functions (displacements), while S-FEM converts further differentials to summations of the values of the field functions. Both weak and W2 formulations have a natural means to handle Neumann (stress) boundary conditions.

Hyperbolic PDEs, on the other hand, are typically solved using FDM, which is simple and effective when used with structured grids with cells and nodes. The Finite Volume Method [7] works well with unstructured grids, using the control volumes to ensure conservation laws, which can be viewed as the integral version of the N–S equations. These methods typically use an Eulerian frame, meaning that the nodes are not moving, and the material flows over the grids. There are methods that solve hyperbolic PDEs but use the Lagrangian frame, in particular the Smoothed Particle Hydrodynamics (SPH).

GSMs work with both Lagrangian and Eulerian meshes/grids, and for deformed solids, fluids, and flowing deformed solids, as discussed in the later chapters. When GSM is used in a Lagrangian mesh, it is termed as L-GSM, just for convenience in discussion.

1.4 Types of numerical methods

1.4.1 *Strong-form vs weak-form governing equations*

There are two main approaches to converting PDEs into SAEs using numerical methods: converting the integral form (weak-form) of PDEs into SAEs, as done in FEM and FVM, or converting the differential form (strong-form) of PDEs directly, as done in FDM. Both approaches reduce the order of PDEs to solvable SAEs.

The numerical methods based on integral-form governing equations have the advantage of holding conservation laws, while the numerical methods based on differential-form governing equations have advantages of simplicity. The computational load and data storage requirement of strong-form methods are relatively lower. Therefore, in many scenarios, especially CFD, strong-form methods are often used.

The technique used for derivative approximation to solve governing PDEs directly is crucial, in terms of accuracy, stability, and efficiency. This becomes even more critical when unstructured mesh/grids are used. The function approximation technique has been known in the past, especially in the development of FEM and meshfree methods. It has been known rather recently that a proper approximation of the gradient of a function is as essential, if not more, in developing effective numerical schemes [1].

The GSM uses the robust derivative approximation technique, known as gradient smoothing, which is an orthogonal projector in W2 formulations as proven by [1]. This enables the GSM to work well with unstructured grids.

1.4.2 *Lagrangian methods vs Eulerian methods*

As discussed earlier, there exist two major fundamental frames for describing governing PDEs: the **Eulerian** description and the **Lagrangian** description. The Eulerian description is a *spatial description*, which is often used in the finite difference method (FDM) [8, 9]. In the Eulerian description, the grid/mesh structure is fixed spatially and does not move, while the media move across the fixed mesh. On the other hand, the Lagrangian description is a *material description* and is often used in the finite element method (FEM) [2]. The grid structure for the numerical methods based on the Lagrangian description is attached to the material. It moves along with the movement and deformation of the material in the computation process. The Lagrangian method works like following spotlights on a stage, tracking each node/element/particle, while the Eulerian methods work like a fixed surveillance camera, recording the passing through node/element/particle.

The numerical methods based on the Lagrangian description can be further classified into two categories (see Figure 1.6): **grid-based methods**

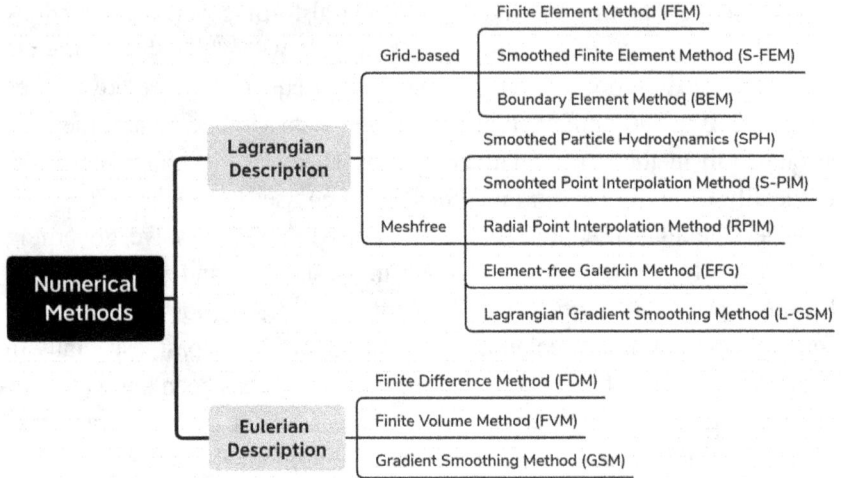

Figure 1.6. Schematic classifications of some of the existing numerical methods (not inclusive).

and **meshfree methods** (including meshless methods and particle methods). Grid-based methods rely more on the quality of the mesh, while meshfree methods place lesser demand on mesh quality. A rough classification of some of the commonly used numerical methods is presented in Figure 1.6. A comprehensive and precise classification of all the existing numerical methods is not an easy task, because of the rich developments over the past century and the fact that some of the methods use operations with mixed features. In addition, the boundary for the classifications is often not clear. Detailed discussion on this can be found in [1].

Taking the most widely used FEM and FDM, respectively, as representatives of the Lagrangian and Eulerian methods, the two types of methods can be compared from the following aspects:

- *On governing equations*: The FEM is built on the basis of weak-form (integral form as discussed in [1]) equations, while FDM is based on strong-form (PDE) equations. Generally, discretization of the strong-form governing equations in FDM is relatively more straightforward to implement, because it uses essentially the Taylor series expansion, and no integration is needed. However, the problem domain needs to be generally regular, and some kinds of structured grids are required, as

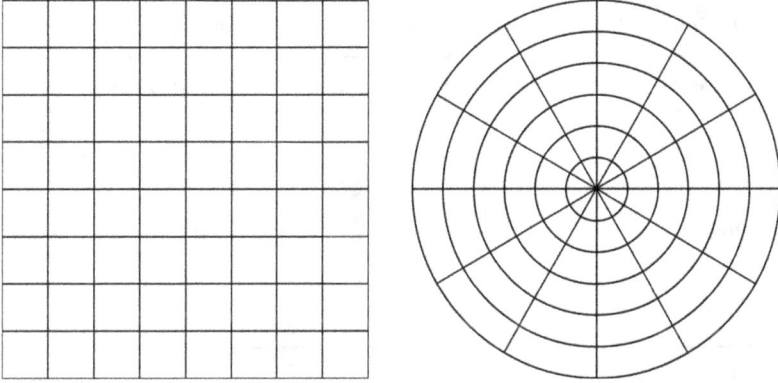

Figure 1.7. Typical structured grids for Eulerian methods based on the Taylor series expansion in the Cartesian coordinates system (left) and Polar coordinates system (right).

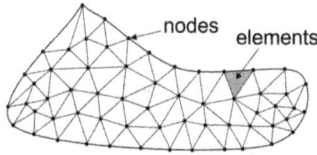

Figure 1.8. Typical unstructured grid for Lagrangian methods.

shown in Figure 1.7. On the contrary, discretization for the weak-form governing equation can use irregular elements for irregular domains, as shown in Figure 1.8. The implementation procedure is, however, more complicated because of the need for numerical integrations in a proper manner [1].

- *On convection terms*: The Eulerian and Lagrangian methods describe motions and flows in a distinct fashion. Because the Eulerian mesh is fixed in space, additional terms in the PDE are needed to account for the convection effects caused by the moving media. On the other hand, the Lagrangian mesh moves with the media, and there is no convection term (caused by the relative motion of the fluid and the mesh) in the PDEs. The differences of the Navier–Stokes equations described under the two frameworks are shown in Table 1.1.
- *On representation of material/media*: In the Eulerian methods, the material is discretized into finite nodes associated with field variables, e.g.,

Table 1.1. Governing PDEs in Lagrangian and Eulerian descriptions [10].

Conservation laws	Lagrangian description	Eulerian description
Mass	$\dfrac{D\rho}{Dt} = -\rho\dfrac{\partial v^\beta}{\partial x^\beta}$	$\dfrac{\partial\rho}{\partial t} + v^\beta\dfrac{\partial\rho}{\partial x^\beta} = -\rho\dfrac{\partial v^\beta}{\partial x^\beta}$
Momentum	$\dfrac{Dv^\beta}{Dt} = -\dfrac{1}{\rho}\dfrac{\partial p}{\partial x^\beta}$	$\dfrac{\partial v^\beta}{\partial t} + v^\alpha\dfrac{\partial v^\beta}{\partial x^\alpha} = -\dfrac{1}{\rho}\dfrac{\partial p}{\partial x^\beta}$
Energy	$\dfrac{De}{Dt} = -\dfrac{p}{\rho}\dfrac{\partial v^\beta}{\partial x^\beta}$	$\dfrac{\partial e}{\partial t} + v^\beta\dfrac{\partial e}{\partial x^\beta} = -\dfrac{p}{\rho}\dfrac{\partial v^\beta}{\partial x^\beta}$

where ρ, e, v, x, and t are density, internal energy, velocity, location, and time, respectively. The Greek superscripts α and β are used to denote the coordinate directions, while the summation is taken over repeated indices. $\partial/\partial t$ and D/Dt refer to the partial derivative and total time derivative (or material derivative), respectively, with the relationship of

$$\frac{D}{Dt} = \frac{\partial}{\partial t} + v^\alpha\frac{\partial}{\partial x^\alpha}$$

This difference gives the additional term as shown in the 3rd column of Table 1.1.

density, pressure, and velocity, whereas the Lagrangian grid-based methods track the mass, volume, deformation, stress, momentum, and energy of the material. This makes the Lagrangian grid-based method preferable for solids, when tracking the movement of the materials, time history of deformation, and stress are essential for analysis.

The Lagrangian grid-based methods are in general superior to Eulerian methods for tracking the history and trajectory of variable fields at each element of the system. However, severe mesh distortion may occur in large deformation problems, which can lead to inaccuracies of numerical solution. This is why Lagrangian grid-based methods cannot, in general, deal with the problems with too extreme movements and large deformation.

1.4.3 *Grid-based methods vs meshfree methods*

Meshfree methods have gained popularity in recent years due to their ability to handle large deformation problems. The main difference between meshfree methods and grid-based methods lies in function approximation strategies and types of weak forms [1]. Grid-based methods (such as the

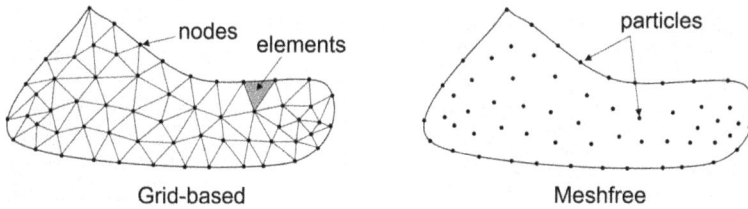

Figure 1.9. Comparison of representations of material or media in Lagrangian grid-based methods and meshfree methods.

FEM) rely on elements, making them sensitive to distorted element shapes, while, for meshfree methods, it is quite difficult to provide universally applicable comments due to their diversity in formulations and operations. In general, they are relatively less dependent on mesh quality and much more tolerant to mesh distortion (see Figure 1.9). This is because meshfree methods often use nodes instead of elements in their formulations and operations or use multiple cells in the background mesh. While some meshfree methods still require a type of mesh, they use only those that can be automatically generated [11], reducing the burden of mesh generation.

The overall comparison of grid-based methods and meshfree methods is listed in Table 1.2.

Substantial efforts have been made in recent decades to develop advanced numerical methods, including meshfree methods, to overcome the limitations of grid-based methods in handling extremely large deformation problems. Many techniques have been found with good performance, such as the element-free Galerkin method [12], Smoothed Particle Hydrodynamics (SPH) [10], and recently the smoothed point interpolation methods (S-PIM) [11, 13–18] and the smoothed radial point interpolation methods (S-RPIM) [19–23], to name just a few. A more comprehensive record on these can be found in books [1, 6, 24] and recent review articles [6, 25–28]. In this book, we will not discuss these general meshfree methods further. The next section will discuss more about some meshfree particle methods, especially the SPH, because it will be used frequently in this book in comparison studies, and it has some commonalities with the GSMs, which are the focus of this book.

Table 1.2.   Comparison of grid-based methods and meshfree methods.

	Eulerian grid-based methods	Lagrangian grid-based methods	Lagrangian Meshfree methods
Grid	Fixed in the space	Elements attached on moving material	Nodes or particle attached on moving material
Track	Mass, momentum, energy flux across cell boundary	Movements of any point on materials	Movements of points on materials
Applicability to extreme movements like flows	Excellent	Good	Very good
Adaptability to complex geometry	Fair	Very good	Excellent
Time history track on material points	Fair	Excellent	Excellent
Interface track	Fair	Very good	Excellent
Stability	Excellent	Excellent	Excellent, but depends on the method
Efficiency	Excellent	Excellent	Excellent, but depends on the method
Implementation	Easy	Easy	Easy, but depends on the method

1.5 Meshfree particle methods

1.5.1 *Concepts of meshfree particle methods*

Meshfree particle methods refer to a class of meshfree methods that employ a set of discrete particles to represent the media or material. Unlike traditional mesh-based methods, meshfree particle methods do not require a mesh in the process of computation. Instead, particles carrying field variables of the problem interact with their neighbors using a smooth function in a local support domain, allowing for the simulation of complex geometries and large deformation.

Meshfree particle methods have the following advantages. First, they possess a simple algorithm and are easy to implement due to the "meshfree" feature. Second, they do not require convective terms in the governing equations, because of the use of Lagrangian frame. Third, they allow for easy tracking of field variables at any material point over time since the particles themselves represent the moving material. Finally, they automatically impose and determine free surfaces, moving boundaries, and material interface.

1.5.2 *A brief theory on SPH*

SPH is the earliest and most popular meshfree particle method, initially developed for discrete astrophysical systems [29, 30]. Because of its outstanding adaptability to extremely large deformation problems, it has been successfully applied to a range of applications, such as blood flows [31, 33, 34], molecular dynamics in multiple scales [35, 36], explosion and underwater shocks [37–39], cracking and fragmentation [40–42], penetration [43–45], high-velocity impact [46–51], and geophysics [52–55]. The basics of SPH can be found in the widely used book [10].

As a meshfree, Lagrangian particle method, SPH has the following features:

1. *Meshfree*: The problem domain is represented by a finite set of arbitrarily distributed particles, each of which carries field variables such as its coordinates, density, velocity, energy, acceleration, and stresses. An SPH particle searches locally for its neighboring particles on the fly to establish their relationship.
2. *Lagrangian*: The dynamics of SPH particles are governed by the strong form of PDEs of N–S equations in the Lagrangian form. There is no convection term and hence no stability issue related to the convection term.
3. *Compact support*: Each SPH particle locally influences its neighboring particles, known as supporting particles. The evolution of field variables of a particle is thus influenced only by its supporting particles.
4. *Adaptive*: The supporting particles' set is updated at every time step, and therefore the involvement of a particle depends on the current local distribution of the neighboring particles.
5. *Explicit*: The dependence of the N–S equations is solved using an explicit time-stepping scheme. The updates for the field variables are

obtained using the following equations for a time increment.

$$\begin{cases} \dfrac{D\rho}{Dt} = -\rho \dfrac{\partial v^\beta}{\partial x^\beta} \\[2mm] \dfrac{Dv^\alpha}{Dt} = \dfrac{1}{\rho} \dfrac{\partial \sigma^{\alpha\beta}}{\partial x^\beta} + g^\alpha \\[2mm] \dfrac{Dx^\alpha}{Dt} = v^\alpha \end{cases} \quad (1.5)$$

where

- ρ, σ, x, and v are the scalar density, total stress tensor, spatial coordinates, and velocity, respectively.
- g^α is a component of the gravity, or any external acceleration.

In the SPH method, the gradient of variables in the N–S equations of a certain SPH particle is approximated using its supporting particles. This is done *via* two key steps: **integral representation** of field functions and **particle approximation**.

The formulation for the **first** step is given as follows. Consider any field function of $f(\mathbf{x})$ with an independent variable $\mathbf{x}$. The gradient of the function ∇f at a point $\mathbf{x}$ is approximated by a weighted integral of ∇f as follows:

$$\langle \nabla f(\mathbf{x}) \rangle = \int_\Omega [\nabla f(\mathbf{x}')] W(\mathbf{x} - \mathbf{x}', h) d\mathbf{x}' \quad (1.6)$$

where W is a locally defined weight function in the local supporting domain Ω of $\mathbf{x}$. It is also known as the *smoothing function* or *kernel function*. The supporting domain Ω is also called *smoothing domain*. The local supporting domain Ω is often a circular area in 2D or a spherical volume in 3D with a radius of h scaled by κ. $d\mathbf{x}'$ is an integral variable representing an infinitesimal volume $d\Omega$. The use of $d\mathbf{x}'$ in Eq. (1.6) may be a little abusive, but it helps to identify the internal variables that vanish after the integration. A typical smoothing domain is shown in Figure 1.10. Note that, in Eq. (1.6),∇f is a vector and hence its multiplication with W is done element wise, meaning each component of the vector is weighted with the same function.

The supporting domain in this case is a circular area with a radius of κh, where κ and h are tunable parameters. The careful choices of these

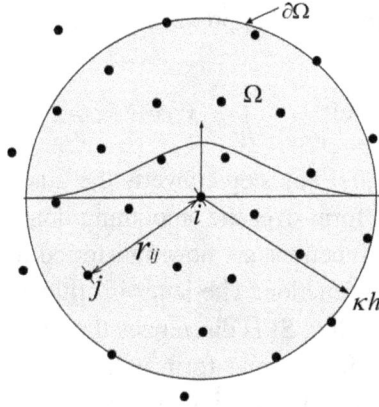

Figure 1.10. Particle approximations using particles located within the supporting domain Ω, where the smoothing function W for particle i is defined

parameters were studied by many and can be found in the recent work by Mao *et al.* [56].

Equation (1.6) is also called *kernel approximation*. In the SPH convention, the kernel approximation of a function is marked by the angle brackets. Using integration by parts, we have

$$[\nabla f(\mathbf{x}')]W(\mathbf{x}-\mathbf{x}',h) = \nabla[f(\mathbf{x}')W(\mathbf{x}-\mathbf{x}',h)] - f(\mathbf{x}')\nabla W(\mathbf{x}-\mathbf{x}',h) \quad (1.7)$$

Substituting the above equation into Eq. (1.6) yields

$$\langle \nabla f(\mathbf{x}) \rangle = \int_{\Omega} \nabla[f(\mathbf{x}')W(\mathbf{x}-\mathbf{x}',h)]d\mathbf{x}' - \int_{\Omega} f(\mathbf{x}')\nabla W(\mathbf{x}-\mathbf{x}',h)d\mathbf{x}' \quad (1.8)$$

The first integral on the right-hand side of Eq. (1.8) can be converted into an integral over the surface of domain Ω:

$$\langle \nabla f(\mathbf{x}) \rangle = \int_{\partial\Omega} f(\mathbf{x}')W(\mathbf{x}-\mathbf{x}',h)\bar{n}dS - \int_{\Omega} f(\mathbf{x}')\nabla W(\mathbf{x}-\mathbf{x}',h)d\mathbf{x}' \quad (1.9)$$

where $\bar{n}$ is the unit out-normal vector to the domain surface $\partial\Omega$. The foregoing equation is in fact the result of Gauss's divergence theorem. Note that the smoothing function needs to be differentiable.

In SPH, the construction of kernel function W is one of the tricky steps, which was also studied by many, including [10], because the smoothing function W is compact, meaning the value of W vanishes on the boundary. Therefore, the surface integral on the right-hand side of Eq. (1.9) become zero. Consequently, the kernel approximation for a spatial derivative has a final form of

$$\langle \nabla f(\mathbf{x}) \rangle = - \int_\Omega f(\mathbf{x}') \nabla W(\mathbf{x} - \mathbf{x}', h) d\mathbf{x}' \qquad (1.10)$$

We note that (1) the *first* key step converts the gradient's calculation into a continuous integral form over the smoothing domain and (2) the differentiations to the field functions are now transferred into differentiations on the known smoothing function. The latter is critical to the stability of the SPH. As mentioned earlier, SPH discretizes the strong-form PDEs of N–S equations. The discretized strong form will be in general unstable [28], but the SPH model is often found rather stable. This is because of the use of Eq. (1.10), which has a "weak-form" feature: The differentiations on an unknown field variable are shifted to the known smoothing function (which is sufficiently smooth by design). As a result, when the 2nd order N-S equation is approximated, we did not really perform differentiations to the field function: The consistency requirement on the field functions is "weakened". This analysis was performed first systematically by GR Liu [1, 11].

In the *second* key step, the continuous integral representation in Eq. (1.10) is discretized as a summation over all the neighboring particles of the target particle. This approximation process is carried out under an assumption, *i.e.*, each SPH particle possesses its own mass and density, and occupies individual volume. The particle volume is thus simply determined by the mass divided by the density obtained from the governing equations in Eq. (1.5). The infinitesimal volume $d\mathbf{x}'$ in Eq. (1.10) can now be replaced with m/ρ, and thus the integral is approximated with a summation over the neighboring particles within the supporting domain:

$$\langle \nabla f(\mathbf{x}_i) \rangle = - \sum_{j=1}^{N} \frac{m_j}{\rho_j} f(\mathbf{x}_j) \nabla_i W(\mathbf{x}_i - \mathbf{x}_j, h) \qquad (1.11)$$

This is the standard gradient approximation formulation of SPH. Similarly, the SPH approximation for an arbitrary function $f(x_i)$ at x_i based on the neighboring particles x_j has a form of

$$\langle f(\mathbf{x}_i) \rangle = \sum_{j=1}^{N} \frac{m_j}{\rho_j} f(\mathbf{x}_j) W(\mathbf{x}_i - \mathbf{x}_j, h) \qquad (1.12)$$

The above particle approximation treatment effectively converts the continuous integral representations of functions and their derivatives into a discretized form with some simple algebraic works. Differential equations become algebraic equations.

Since the supporting particles are updated at every time step, this adaptable nature enables the SPH method to handle large deformation problems naturally. This is an attractive feature of the SPH method.

1.5.3 *Pros and Cons of the SPH method*

Compared to the other meshfree particle methods, the SPH method has the following practical features in applications:

- *Good adaptability* to extremely large deformation problems. The gradient of field variables in governing PDEs can be approximated rather easily and quite robustly for reasonable numerical stability for highly dynamic problems.
- *Easy implementation*: SPH is a particle method where the problem domain is represented by a finite number of SPH particles connecting to its neighbors without using a mesh during computations.
- *Easy parallelization*: Considering the explicit time integration and the independent gradient approximation of field variables on each SPH particle, parallel computing algorithms can be developed for fast computations.

However, the SPH method does suffer from some inherent drawbacks:

- *Tensile instability*: It has been frequently shown that the SPH method becomes unstable as particles are subjected to tensile stress. This phenomenon is well known as the *tensile instability* in SPH [10, 54, 57–61]. Although corrective solutions have been developed, e.g., by introducing additional artificial stress [57, 59] or through adaptively choosing kernel function for each particle based on its stress state [62, 63], the additional corrections may lead to alternative issues.
- *Low efficiency in computation*: This is because of the need to update the kernel function and neighboring particle sets at every time step, which can be quite time-consuming.

It is a challenge to develop a numerical method that possesses all qualities of *accuracy, stability, efficiency, adaptability, and ease in implementation.* An optimal balance needs to be achieved.

1.6　Gradient smoothing methods

The GSMs are a relatively new type of computational methods based on a robust and adaptable gradient smoothing technique for approximating derivatives accurately over an unstructured grid. The most important merit of the GSMs is their excellent applicability to problems with an extremely distorted mesh or distributed particles.

The GSMs use techniques from SPH and the smoothed finite element method [5]. They also have some similarities with the Finite Volume Method (FVM) [64]. The GSMs essentially use Eq. (1.9), but only the first term on the right-hand side, and discard the second term by properly choosing the smoothing function. This is different from SPH, because SPH does the opposite: discard the first term, but keep the second term, via different smoothing functions. This gradient smoothing is in fact an orthogonal projector as proven in W2 formulations by [1].

The GSMs were formulated initially with the Eulerian description [65–70], which is named as GSM. Recently, they were also formulated in the Lagrangian frame, called "Lagrangian Gradient Smoothing Method" or L-GSM [71–75]. The GSM and L-GSM together are referred to as "Gradient Smoothing Methods" in this book.

1.6.1　*Brief on GSM*

Motivations of GSM: It has been stated in Section 1.4.2 that the main limitation of the widely used FDM is the need for a structured grid to approximate the derivatives, leading to difficulty for the problems with a complex geometric shape. The "GSM" was proposed to remove this limitation by accurately approximating derivatives over an unstructured grid, as shown in Figure 1.11.

FDM vs FVM: FDM and FVM are two commonly used numerical methods in Computational Fluid Dynamics (CFD). FDM discretizes the domain into finite "nodes" and approximates the derivatives of field variables at each

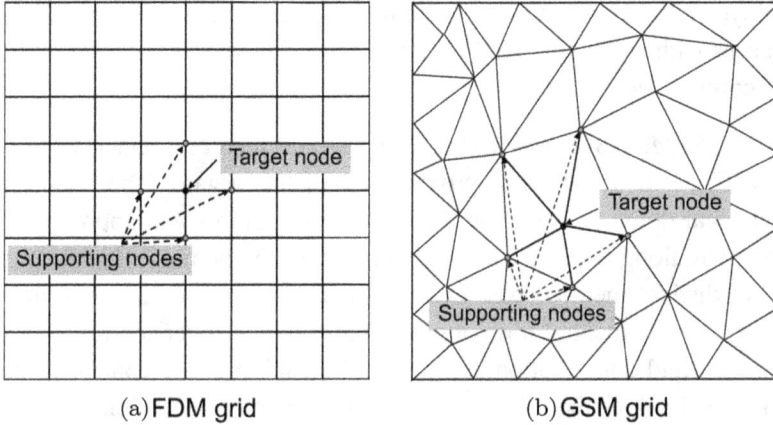

(a) FDM grid   (b) GSM grid

Figure 1.11. Comparison of grid structures for FDM and GSM.

node with the Tylor expansion theory based on local information. FVM discretizes the domain into finite "control volumes" and approximates the field variables' integral over these control volumes following the conservation laws of transport across control volumes. So, FDM and FVM are totally different methods in this regard. FDM has two main advantages over FVM. One is a much easier implementation procedure: FVM needs to construct a "control volume" while FDM only demands an imaginary structured mesh and is thus free from complicated mesh construction. The other advantage is in boundary treatment as a strong-form method over the FVM solving weak-form governing equations. FVM has its own advantage over FDM. Since FDM relies on a regular distribution of nodes, it cannot handle those problems with complex geometries or boundaries. In this regard, FVM works well because its control volume can be constructed without any shape constraint. In sum, FDM can be a preferable option for those problems with regular geometries and boundaries while FVM will be preferable for problems with complex shapes.

GSM vs FDM: GSM solves the same governing equations in Eulerian description as FDM, and therefore has an FDM-like numerical framework. The main difference lies in the gradient approximating strategies and mesh implementation, which will be presented in detail in Chapter 3. It has been shown that the GSM becomes the FDM when a uniform grid is employed [71]. Hence, GSM can be regarded as the extension of FDM for the cases with an unstructured grid. It is thus capable of handling problems with complex geometry.

GSM vs FVM: GSM uses Gauss's divergence theorem, and it is similar to the FVM in derivative approximation in this regard. They both construct local domains based on the surrounding points and approximating derivatives along the edge/surface boundary of the domain. However, they are two distinct methods based on totally different discretization fashions. FVM solves the integral form of governing equations over control volumes, requiring rigid conservation and physical meaning of the control volume, while GSM solves the differential form of governing equations and holds the conservation law in a differential manner. GSM applies to any PDEs regardless of whatever conservation conditions. In addition, GSM works in both Eulerian and Lagrangian frames.

GSM has good capability of handling complex geometries similar to FVM. It is a general numerical method for discretizing all types of PDEs including elliptic solid mechanics problems [69], hyperbolic flows, and flowing solids in Euler or Lagrangian descriptions [76, 77].

1.6.2 *Brief on L-GSM*

Motivations of L-GSM: As mentioned in Section 1.5.3, the SPH method is advantageous in handling large deformation problems but suffers from the "tensile instability" issue and low computational efficiency. In particular, the inherent instability issue is increasingly becoming the primary bottleneck of SPH. A study conducted by the senior author's group on the node-based smoothed point interpolation method (NS-PIM) [14, 15, 78] and the node-based smoothed finite element method [79–81] found that node-based models can exhibit instability for dynamics problems, even though their spatial stability has theoretically been proven. Although the NS-PIM and NS-FEM are based on weakened weak form (W2), the lack of sampling (caused by assuming the gradient is a constant in a smoothing domain) for gradient approximation seemed to be the root of the instability. Based on this discovery, the senior author suspected that the instability of SPH may also be rooted at insufficient particle-based sampling (caused by simple particle summation for all particles in a smoothing domain). Also, SPH only utilizes the location information of neighboring particles relative to the target particle, while the relative location information among the neighboring particles are ignored. Therefore, the gradient approximation techniques used in the GSM may be able to resolve the instability issue of SPH from the root cause. This is because gradient approximation in GSM is smoothed domain based, rather than based on an entire set of all supporting particles in the SPH. To equip GSM with other the good features of SPH, we formulated GSM with the Lagrangian description, which led to the L-GSM.

In addition, the low efficiency of SPH is mainly caused by the necessary time-consuming searching process to support particles at every time step. Since the connection of L-GSM particles is based on a background grid (see Figure 1.12), it does not need to be renewed frequently. This feature makes L-GSM more efficient in computation than SPH under the same conditions, especially when a large number of particles are employed in the simulation.

The L-GSM is, thus, a kind of SPH-like particle method but with GSM gradient approximation strategies. Hence, they are similar in many aspects of numerical implementation. The difference is only the gradient approximation strategy and the use of different types of smoothing functions. SPH uses a bell-shaped function that vanishes on the boundary of the smoothing domain, but GSM uses Heaviside step functions whose gradient vanishes inside the smoothing domain. Changes are thus needed in establishing the connection of particles *via* construction of a supporting domain or a smoothing domain. Details of the L-GSM theory will be presented in Chapter 6.

This book is based on thesis work done by Xu [82] on GSM and Mao [83] on L-GSM, under the supervision of the senior author.

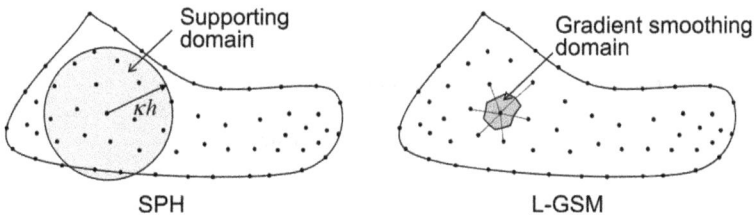

Figure 1.12. Supporting domain of SPH and gradient approximations in L-GSM.

1.7 Notation

This section defines the major notations used in this book. Both the widely used matrix/vector notations and indicial notations will be used for convenience of description and ease of comprehension.

Vector and matrix: The quantities of vector, and the matrix, will be in boldface.

Subscripts and superscripts: The *subscripts* and *superscripts* are all written in *Italic* fonts.

Operators of summation and product: The notation

$$\sum_i f(i)$$

will denote a summation of all the function values indexed by i.

Symbols	Physics meanings
E	Yong's modulus
E_t	Total internal energy
$\mathbf{F}_c$	Vector of convective fluxes
$\mathbf{F}_{c,T}$	Turbulent convective fluxes
$\mathbf{F}_v$	Vector of viscous fluxes
$\mathbf{F}_{v,T}$	Turbulent viscous fluxes
$\mathbf{F}_{\text{external}}$	External force vector
G	Shear modulus
$\mathbf{I}$	Unit diagonal matrix
I_1	First invariant of stress tensor
J_2	Second invariant of deviatoric stress tensor
K	Elastic bulk modulus
Superscript ^{L,R}	Left, right
Subscript ∞	Reference value
Ma	Mach number
Pr	Prandtl number
Pr_T	Turbulent Prandtl number
R	Specific gas constant
Re	Reynolds number
S	Projected area of smoothing domain
S_T	Source term in turbulence model
St	Strouhal number
T	Temperature

(Continued)

(Continued)

Symbols	Physics meanings
T_p	Period time
$U(\mathbf{x})$	An arbitrary scalar function in terms of coordinate $\mathbf{x}$
V	Volume of gradient smoothing domain
W	Weight or smoothing function
$\mathbf{d}$	Displacement vector
c	Cohesion of soil material
c_s	Sound of speed
c_p	Specific heat at constant pressure
e	Interval energy
$\dot{\mathbf{e}}$	Deviatoric shear strain rate tensor
$\mathbf{g}$	Gravity acceleration
g_p	Plastic potential
h	Smoothing length or spacing size
k_{eff}	Effective thermal conductivity
k_T	Turbulent thermal conductivity
m	Mass
$\bar{n}$ or $\mathbf{n}$	Normal unit vector of curve or curved surface
n_x, n_y	Components of in x and y directions, respectively
p	Pressure
p_{wall}	Pressure of wall
r	Length of edge or radius
$\mathbf{r}_D$	Directional vector
$\mathbf{r}_Q$	Eigenvector of matrix
t	Time
$\mathbf{t}$	Unit tangent vector
u	Horizontal velocity
v	Vertical velocity
$\tilde{v}$	Kinematic turbulent viscosity
$\mathbf{v}$	Velocity vector
$\mathbf{x}$	Coordinate vector (x, y, z)
Ψ	Stream function
Ω	Gradient smoothing domain
Π	Artificial viscosity in SPH and L-GSM
α_Π, β_Π	Parameters in artificial viscosity
β	Artificial compressibility
$\dot{\boldsymbol{\varepsilon}}$	Total strain rate vector
$\dot{\boldsymbol{\varepsilon}}^e$	Elastic strain rate vector
$\dot{\boldsymbol{\varepsilon}}^p$	Plastic strain rate vector
ϕ	Internal friction angle of soil
γ	Ratio of specific heat of fluid media

(Continued)

(*Continued*)

Symbols	Physics meanings
κ	Scaling factor of smoothing length in SPH
λ	Secondary viscous coefficient
μ	Dynamic viscosity coefficient
μ_{eff}	Effective dynamic viscosity
μ_T	Turbulent dynamic viscosity
v	Poisson's ratio
ν	Kinematic viscosity
ρ	Density
σ	Total stress tensor
$\dot{\sigma}$	Stress rate tensor
$\tau_{xx}, \tau_{xy}, \tau_{yy}$	Components of viscous stress tensor
τ	Deviatoric stress
τ	Pseudo time
ω	Vorticity
$\dot{\omega}$	Spin rate tensor
ψ	Dilatancy angle
ε_{con}	Convergence error index
$\partial\Omega$	Boundary of support domain
δ	Kronecker delta
Δt	Physical time step
$\Delta\tau$	Pseudo time step

References

[1] G. R. Liu, *Meshfree Methods: Moving Beyond the Finite Element Method*. Boca Raton, FL: CRC Press (2010).

[2] G. R. Liu, *Finite Element Method: A Practical Course*. Oxford: Butterworth-Heinemann (2003).

[3] K. Poochinapan, -H. Hsu, M. Langthjem, and M. Mei, "Numerical Implementations for 2D Lid-Driven Cavity Flow in Stream Function Formulation," *International Scholarly Research Network ISRN Applied Mathematics*, 2012, 17 (2012). doi: 10.5402/2012/871538.

[4] G. R. Liu, and X. Han, "Computational inverse techniques in nondestructive evaluation." In *Computational Inverse Techniques in Nondestructive Evaluation*, CRC Press, pp. 1–567 (2003, January). doi: 10.1201/9780203494486.

[5] G. R. Liu and T. Trung. Nguyen, *Smoothed Finite Element Methods*. Boca Raton, FL: CRC Press (2010).

[6] G. R. Liu, "An Overview on Meshfree Methods: For Computational Solid Mechanics," *International Journal of Computational Methods*, 13, 5 (2016, October). doi: 10.1142/S0219876216300014.

[7] R. J. LeVeque, "Finite Volume Methods for Hyperbolic Problems," *Finite Volume Methods for Hyperbolic Problems* (2002, August). doi: 10.1017/CBO978051179 1253.

[8] G. D. Smith, *Numerical Solution of Partial Differential Equations: Finite Difference Methods*. Clarendon Press (1985).

[9] J. W. Thomas, *Numerical Partial Differential Equations: Finite Difference Methods*, vol. 22. New York, NY: Springer (1995). doi: 10.1007/978-1-4899-7278-1.

[10] G. R. Liu and M. B. Liu, *Smoothed Particle Hydrodynamics: A Meshfree Particle Method*. Singapore: World Scientific (2003).

[11] G. R. Liu and G. Y. Zhang, *Smoothed Point Interpolation Methods: G Space Theory and Weakened Weak Forms*. Singapore: World Scientific Publishing Co., (2013). doi: 10.1142/8742.

[12] T. Belytschko, Y. Y. Lu, and L. Gu, "Element-free Galerkin methods," *International Journal for Numerical Methods in Engineering*, 37 (2), 229–256 (1994, January). doi: 10.1002/NME.1620370205.

[13] G. R. Liu and G. Y. Zhang, "Edge-based smoothed point interpolation methods," *International Journal of Computational Methods*, 5(4), 621–646 (2008, November). doi: 10.1142/S0219876208001662.

[14] S. C. Wu, G. R. Liu, H. O. Zhang, and G. Y. Zhang, "A node-based smoothed point interpolation method (NS-PIM) for three-dimensional thermoelastic problems," *Numerical Heat Transfer; Part A: Applications*, 54(2), 1121–1147 (2008, January). doi: 10.1080/10407780802483516.

[15] Q. Tang, G. Y. Zhang, G. R. Liu, Z. H. Zhong, and Z. C. He, "A three-dimensional adaptive analysis using the meshfree node-based smoothed point interpolation method (NS-PIM)," *Engineering Analysis with Boundary Elements*, 35(10), 1123–1135 (2011, October). doi: 10.1016/j.enganabound.2010.05.019.

[16] G. Y. Zhang and G. R. Liu, "Meshfree cell-based smoothed point interpolation method using isoparametric pim shape functions and condensed rpim shape functions," *International Journal of Computational Methods*, 8(4), 705–730 (2011, December). doi: 10.1142/S0219876211002770.

[17] G. R. Liu and G. Y. Zhang, "A normed G space and weakened weak (W2) formulation of a cell-based smoothed point interpolation method," *International Journal of Computational Methods*, 6(1), 147–179 (2009, November). doi: 10.1142/S0219876209001796.

[18] A. Tootoonchi, A. Khoshghalb, and G. R. Liu, "A novel approach for application of smoothed point interpolation methods to axisymmetric problems in poroelasticity," *Computers and Geotechnics*, 102, 39–52 (2018, October). doi: 10.1016/j.compgeo.2018.05.010.

[19] G. R. Liu, Y. Jiang, L. Chen, G. Y. Zhang, and Y. W. Zhang, "A singular cell-based smoothed radial point interpolation method for fracture problems," *Computers & Structures*, 89(13–14), 1378–1396 (2011, July). doi: 10.1016/j.compstruc.2011.03.009.

[20] G. R. Liu, G. Y. Zhang, Z. Zong, and M. Li, "Meshfree cell-based smoothed alpha radial point interpolation method (CS-α RPIM) for solid mechanics problems," *International Journal of Computational Methods*, 10(4), 1350020 (2013, August). doi: 10.1142/S0219876213500205.

[21] Z. C. He, G. Y. Li, E. Li, Z. H. Zhong, and G. R. Liu, "MID-frequency acoustic analysis using edge-based smoothed tetrahedron radialpoint interpolation methods," *International Journal of Computational Methods*, 11(5) 1350103 (2014, October). doi: 10.1142/S021987621350103X.

[22] Y. Li, G. R. Liu, and J. H. Yue, "A novel node-based smoothed radial point interpolation method for 2D and 3D solid mechanics problems," *Computers & Structures*, 196, 157–172 (2018, February). doi: 10.1016/j.compstruc.2017.11.010.

[23] Y. Li and G. R. Liu, "An element-free smoothed radial point interpolation method (EFS-RPIM) for 2D and 3D solid mechanics problems," *Computers & Mathematics with Applications*, 77(2), 441–465 (2019, January). doi: 10.1016/j.camwa.2018.09.047.

[24] G. R. Liu and Y. T. Gu, *An Introduction to Meshfree Methods and Their Programming* (1st ed). Netherlands: Springer Netherlands (2005). Accessed: November 9, 2019. Available: https://www.springer.com/gp/book/9781402032288.

[25] G. R. Liu, "The smoothed finite element method (S-FEM): A framework for the design of numerical models for desired solutions," *Frontiers of Structural and Civil Engineering* , 13(2), 456–477 (2019, April 1). doi: 10.1007/s11709-019-0519-5.

[26] W. Zeng and G. R. Liu, "Smoothed finite element methods (S-FEM): An overview and recent developments," *Archives of Computational Methods in Engineering*, 25(2), 397–435 (2018, April). doi: 10.1007/s11831-016-9202-3.

[27] M. B. Liu, G. R. Liu, L. W. Zhou, and J. Z. Chang, "Dissipative particle dynamics (DPD): An overview and recent developments," *Archives of Computational Methods in Engineering*, 22(4), 529–556 (2015, November). doi: 10.1007/s11831-014-9124-x.

[28] M. B. Liu and G. R. Liu, "Smoothed particle hydrodynamics (SPH): An overview and recent developments," *Archives of Computational Methods in Engineering*, 17(1), 25–76 (2010, March). doi: 10.1007/s11831-010-9040-7.

[29] L. B. Lucy and L. B., "A numerical approach to the testing of the fission hypothesis," *The Astronomical Journal*, 82, 1013–1024 (1977, December).

[30] R. A. Gingold and J. J. Monaghan, "Smoothed particle hydrodynamics-theory and application to non-spherical stars," *Monthly Notices of the Royal Astronomical Society*, 181, 375–389 (1977). doi: 10.1093/mnras/181.3.375.

[31] M. Müller, S. Schirm, and M. Teschner, "Interactive blood simulation for virtual surgery based on smoothed particle hydrodynamics.," *Technology and Health Care*, 12(1) 25–31 (2004).

[32] S. E. Hieber, J. H. Walther, and P. Koumoutsakos, "Remeshed smoothed particle hydrodynamics simulation of the mechanical behavior of human organs," *Technology and Health Care*, 12(4), 305–314 (2004).

[33] N. Tanaka and T. Takano, "Microscopic-scale simulation of blood flow using SPH method," *International Journal of Computational Methods*, 2(4), 555–568 (2005).

[34] K. Tsubota, S. Wada, and T. Yamaguchi, "Simulation study on effects of hematocrit on blood flow properties using particle method," *Journal of Biomechanical Science and Engineering*, 1(1), 159–170 (2006).

[35] X. Yan, Y.-T. Jiang, C.-F. Li, R. R. Martin, and S.-M. Hu, "Multiphase SPH simulation for interactive fluids and solids," *ACM Transactions on Graphics*, 35(4), 1–11 (2016). doi: 10.1145/2897824.2925897.

[36] J. J. Monaghan and A. Kocharyan, "SPH simulation of multi-phase flow," *Computer Physics Communications*, 87(1–2), 225–235 (1995).

[37] M. B. Liu, D. L. Feng, and Z. M. Guo, "Recent developments of SPH in modelling explosion and impact problems," in *International Conference on Paricle-based Methods – Fundamentals and Applications*, 1–8, 428–435 (2013).

[38] M. B. Liu, G. R. Liu, K. Y. Lam, and Z. Zong, "Smoothed particle hydrodynamics for numerical simulation of underwater explosion," *Computational Mechanics*, 30(2), 106–118 (2003, January).

[39] M. B. Liu, G. R. Liu, Z. Zong, and K. Y. Lam, "Computer simulation of high explosive explosion using smoothed particle hydrodynamics methodology," *Computers & Fluids*, 32(3), 305–322 (2003).

[40] T. Rabczuk and J. Eibl, "Simulation of high velocity concrete fragmentation using SPH/MLSPH," *International Journal for Numerical Methods in Engineering*, 56(10), 1421–1444 (2003, March).

[41] T. Rabczuk and T. Belytschko, "A three-dimensional large deformation meshfree method for arbitrary evolving cracks," *Computer Methods in Applied Mechanics and Engineering*, 196(29–30), 2777–2799 (2007, May).

[42] T. Rabczuk, G. Zi, S. Bordas, and H. Nguyen-Xuan, "A simple and robust three-dimensional cracking-particle method without enrichment," *Computer Methods in Applied Mechanics and Engineering*, 199(37–40), 2437–2455 (2010, August).

[43] R. F. Kulak, "Modeling of Cone Penetration Test Using SPH and MM-ALE Approaches," in *8th European LS-DYNA Users Conference*, Strasbourg, 1–10, 55263663 (2011).

[44] L. Ma, R. hao Bao, and Y. mu Guo, "Waterjet penetration simulation by hybrid code of SPH and FEA," *International Journal of Impact Engineering*, 35(9), 1035–1042 (2008).

[45] S. Seo, O. Min, and J. Lee, "Application of an improved contact algorithm for penetration analysis in SPH," *International Journal of Impact Engineering*, 35(6), 578–588 (2008).

[46] L. Aktay and A. F. Johnson, "FEM/SPH coupling technique for high velocity impact simulations," in *Advances in Meshfree Techniques*, pp. 147–167 (Part of the *Computational Methods in Applied Sciences*, Book Series (COMPUTMETHODS, Volume 5) (2007).

[47] K. A. Alhussan, V. A. Babenko, I. M. Kozlov, and A. S. Smetannikov, "Development of modified SPH approach for modeling of high-velocity impact," *International Journal of Heat and Mass Transfer*, 55(23–24), 6340–6348 (2012).

[48] A. Grimaldi, A. Sollo, M. Guida, and F. Marulo, "Parametric study of a SPH high velocity impact analysis - A birdstrike windshield application," *Composite Structures*, 96, 616–630 (2013).

[49] G. R. Johnson, R. A. Stryk, and S. R. Beissel, "SPH for high velocity impact computations," *Computer Methods in Applied Mechanics and Engineering*, 139(1–4), 347–373 (1996).

[50] G. R. Liu, C. E. Zhou, and G. Y. Wang, "An implementation of the smoothed particle hydrodynamics for hypervelocity impacts and penetration to layered composites," *International Journal of Computational Methods*, 10(3), (2013, June). doi: 10.1142/S0219876213500564.

[51] V. Mehra and S. Chaturvedi, "High velocity impact of metal sphere on thin metallic plates: A comparative smooth particle hydrodynamics study," *Journal of Computational Physics*, 212(1), 318–337 (2006).

[52] Z. Dai, Y. Huang, H. Cheng, and Q. Xu, "3D numerical modeling using smoothed particle hydrodynamics of flow-like landslide propagation triggered by the 2008 Wenchuan earthquake," *Engineering Geology*, 180, 21–33 (2014, October).

[53] M. Hu, M. B. Liu, M. W. Xie, and G. R. Liu, "Three-dimensional run-out analysis and prediction of flow-like landslides using smoothed particle hydrodynamics," *Environmental Earth Sciences*, 73(4), 1629–1640 (2015, February).

[54] H. H. Bui, R. Fukagawa, K. Sako, and S. Ohno, "Lagrangian meshfree particles method (SPH) for large deformation and failure flows of geomaterial using elastic-plastic soil constitutive model," *International Journal for Numerical and Analytical Methods in Geomechanics*, 32(12), 1537–1570 (2008, August).

[55] Z. Mao and G. R. Liu, "A smoothed particle hydrodynamics model for electrostatic transport of charged lunar dust on the moon surface," *Computational Particle Mechanics*, 5, 539–551 (2018) doi: 10.1007/s40571-018-0189-4.

[56] Z. Mao, G. R. Liu, and X. Dong, "A comprehensive study on the parameters setting in smoothed particle hydrodynamics (SPH) method applied to hydrodynamics problems," *Computers and Geotechnics*, 92, 77–95 (2017).

[57] J. J. Monaghan, "SPH without a Tensile Instability," *Journal of Computational Physics*, 159(2), 290–311 (2000).

[58] Z. Mao, G. R. Liu, and X. Dong, "A comprehensive study on the parameters setting in smoothed particle hydrodynamics (SPH) method applied to hydrodynamics problems," *Computers and Geotechnics*, 92, 77–95 (2017, December). doi: 10.1016/j.compgeo.2017.07.024.

[59] V. Mehra, C. D. Sijoy, V. Mishra, and S. Chaturvedi, "Tensile Instability and Artificial Stresses in Impact Problems in SPH," *Journal of Physics: Conference Series*, 377, 012102 (2012).

[60] Y. Meleán, L. D. G. Sigalotti, and A. Hasmy, "On the SPH tensile instability in forming viscous liquid drops," *Computer Physics Communications*, 157(3), 191–200 (2004).

[61] J.-Z. Zhang, J. Zheng, K.-P. Yu, and Y.-J. Wei, "A research on the tensile instability of SPH in fluid dynamics," *Gongcheng Lixue/Engineering Mechanics*, 27(2), 65–72 (2010).

[62] F. V. Sirotkin and J. J. Yoh, "A new particle method for simulating breakup of liquid jets," *Journal of Computational Physics*, 231(4), 1650–1674 (2012, February). doi: 10.1016/J.JCP.2011.10.020.

[63] X. Yang, M. Liu, and S. Peng, "Smoothed particle hydrodynamics modeling of viscous liquid drop without tensile instability," *Computers & Fluids*, 92, 199–208 (2014, March). doi: 10.1016/J.COMPFLUID.2014.01.002.

[64] F. Moukalled, L. Mangani, and M. Darwish, *The Finite Volume Method in Computational Fluid Dynamics*, vol. 113, (2016). Cham: Springer. doi: 10.1007/978-3-319-16874-6.

[65] G. R. Liu, *et al.*, "A gradient smoothing method (GSM) with directional correction for solid mechanics problems," *Computational Mechanics*, 41(3), 457–472 (2007, December).

[66] G. R. LIU, "A generalized gradient smoothing technique and the smoothed bilinear form for galerkin formulation of a wide class of computational methods," *International Journal of Computational Methods*, 5(2), 199–236 (2008). doi: 10.1142/S0219876208001510.

[67] G. X. Xu, G. R. Liu, and A. Tani, "An adaptive gradient smoothing method (GSM) for fluid dynamics problems," *International Journal for Numerical Methods in Fluids*, 62(5), 499–529 (2010, February). doi: 10.1002/fld.2032.

[68] J. Yao, G. R. Liu, D. Qian, C. L. Chen, and G. Xu, "A moving-mesh gradient smoothing method for compressible CFD problems," *Mathematical Models and Methods in Applied Sciences*, 23(2), 273–305 (2013, February).

[69] J. Zhang, G. R. Liu, K. Y. Lam, H. Li, and G. Xu, "A gradient smoothing method (GSM) based on strong form governing equation for adaptive analysis of solid mechanics problems," *Finite Elements in Analysis and Design*, 44(15), 889–909 (2008, November).

[70] Y. Chai, X. You, W. Li, Y. Huang, Z. Yue, and M. Wang, "Application of the edge-based gradient smoothing technique to acoustic radiation and acoustic scattering from rigid and elastic structures in two dimensions," *Computers & Structures*, 203, 43–58 (2018, July). doi: 10.1016/J.COMPSTRUC.2018.05.009.

[71] Z. Mao and G. R. Liu, "A Lagrangian gradient smoothing method for solid-flow problems using simplicial mesh," *International Journal for Numerical Methods in Engineering*, 113(5), 858–890 (2018, February). doi: 10.1002/nme.5639.

[72] Z. Mao, G. R. Liu, and Y. Huang, "A local Lagrangian gradient smoothing method for fluids and fluid-like solids: A novel particle-like method," *Engineering Analysis with Boundary Elements*, 107, 96–114 (2019, October). doi: 10.1016/j.enganabound.2019.07.003.

[73] Z. Mao, G. Liu, Y. Huang, and Y. Bao, "A conservative and consistent Lagrangian gradient smoothing method for earthquake-induced landslide simulation," *Engineering Geology*, 260, 105226 (2019, October). doi: 10.1016/j.enggeo.2019.105226.

[74] Z. Mao, G. R. Liu, X. Dong, and T. Lin, "A conservative and consistent Lagrangian gradient smoothing method for simulating free surface flows in hydrodynamics," *Computational Particle Mechanics*, 6, 781–801 (2019). doi: 10.1007/s40571-019-00262-z.

[75] Z. Mao and G. R. Liu, "A 3D Lagrangian gradient smoothing method framework with an adaptable gradient smoothing domain-constructing algorithm for simulating large deformation free surface flows," *International Journal for Numerical Methods in Engineering*, 121(6), 1268–1296 (2020, March). doi: 10.1002/nme.6265.

[76] Z. Mao, G. R. Liu, X. Dong, and T. Lin, "A conservative and consistent Lagrangian gradient smoothing method for simulating free surface flows in hydrodynamics," *Computational Particle Mechanics*, 6(4), (2019). doi: 10.1007/s40571-019-00262-z.

[77] G. R. Liu and G. X. Xu, "A gradient smoothing method (GSM) for fluid dynamics problems," *International Journal for Numerical Methods in Fluids*, 58(10), 1101–1133 (2008, December).

[78] G. R. Liu and G. Y. Zhang, "Upper and Lower Bounds for Numerical Solutions of Elasticity Problems Using LC-PIM and FEM," Springer, Berlin, Heidelberg, pp. 140–155 (2007). doi: 10.1007/978-3-540-75999-7_13.

[79] G. R. Liu, T. Nguyen-Thoi, H. Nguyen-Xuan, and K. Y. Lam, "A node-based smoothed finite element method (NS-FEM) for upper bound solutions to solid mechanics problems," *Computers & Structures*, 87(1–2), 14–26 (2009, January). doi: 10.1016/j.compstruc.2008.09.003.

[80] G. Liu, M. Chen, and M. Li, "Lower bound of vibration modes using the node-based smoothed finite element method (NS-FEM)," *International Journal of Computational Methods*, 14(4), 1750036 (2017, August). doi: 10.1142/S0219876217500360.

[81] T. Nguyen-Thoi, G. R. Liu, and H. Nguyen-Xuan, "Additional properties of the node-based smoothed finite element method (NS-FEM) for solid mechanics problems," *International Journal of Computational Methods*, 6(4), 633–666 (2009, December). doi: 10.1142/S0219876209001954.

[82] X. Xu, "Development of Gradient Smoothing Method (GSM) for fluid flow problems," National University of Singapore (2009). doi: https://scholarbank.nus.edu.sg/handle/10635/160959.

[83] Z. Mao, "A Novel Lagrangian Gradient Smoothing Method for Fluids and Flowing Solids," University of Cincinnati, (2019). Accessed: January, 04, 2021. Available: http://rave.ohiolink.edu/etdc/view?acc_num$=$ucin1553252214052311.

[84] G. R. Liu, *Machine Learning with Python: Theory and Applications*, World Scientific (2023).

Chapter 2

Theory for Gradient Smoothing Methods

Contents

The gradient smoothing method (GSM) is a robust and widely applicable method for approximating gradients in problem domains with unstructured meshes or irregularly distributed nodes or particles. This technique involves constructing a local smoothing domain around the point of interest and using a smoothing function to average the gradient of field variables.

The GSM consists of several components, including a *gradient smoothing operator*, *smoothing functions*, a *gradient smoothing domain (GSD)*, and a *spatial approximation scheme*. The operator calculates the gradient of field variables using the smoothing function and the spatial approximation scheme. The smoothing function determines the weighting of data over the smoothing domain, and the GSD is constructed using neighboring points or particles that are close to the point of interest. The spatial approximation scheme controls the shape of the smoothing domain and the order of the polynomial approximation used to construct the smoothing function.

Based on the type of smoothing functions used, the GSM can be divided into two main categories: constant-weighted GSM and linearly-weighted GSM (LW-GSM). Both will be formulated and examined.

2.1 Constant-weighted gradient smoothing

2.1.1 *Gradient approximation*

Consider a problem of flowing media that occupies a domain Ω and is bounded by $\partial\Omega$. To solve such a problem numerically for field variable functions, we first need to approximate the gradients of these functions at any point $\mathbf{x}$ in a small local domain Ω^s. Similar to the SPH described in Chapter 1, GSM approximates the gradient of a function using also a gradient smoothing operator [1–5] defined in Ω^s. It is an integral representation of the gradient ∇U over Ω^s:

$$\nabla U(\mathbf{x}) = \int_{\Omega^s} \nabla U(\mathbf{x}')W(\mathbf{x} - \mathbf{x}')d\mathbf{x}' \qquad (2.1)$$

where Ω^s is called the smoothing domain and W is the smoothing function. It is often denoted by W because it is also called the weight function in the SPH literature. In general, the smoothing function W can have various forms and should be at least piecewise differentiable in Ω^s.

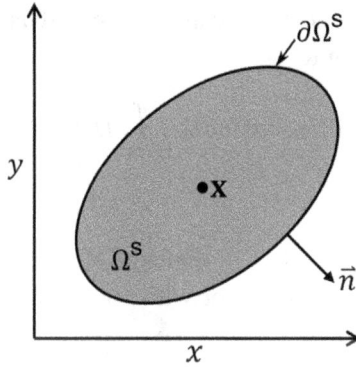

Figure 2.1. Sketch of a smoothing domain Ω^s in 2D for gradient approximation of a function at a point of interest **x**.

The smoothing domain Ω^s in GSM corresponds to the supporting domain in SPH, and it can have an arbitrary shape covering the point of interest **x**. In GSM implementations, the smoothing domain is often a polygon, but does not have to. Figure 2.1 shows a simple elliptical smoothing domain defined in a two-dimensional (2D) space.

The accuracy of Eq. (2.1) depends on the choice of smoothing function W. Detailed requirements on W are discussed in [6]. The most basic requirement is that a *constant* gradient of U should be exactly reproduced, which requires

$$\nabla U = \int_{\Omega^s} \underbrace{\nabla U(\mathbf{x}')}_{\mathbf{C}} W(\mathbf{x} - \mathbf{x}')d\mathbf{x}' = \int_{\Omega^s} \mathbf{C}W(\mathbf{x} - \mathbf{x}')d\mathbf{x}'$$

$$= \mathbf{C}\int_{\Omega^s} W(\mathbf{x} - \mathbf{x}')d\mathbf{x}' \equiv \mathbf{C} \qquad (2.2)$$

This means

$$\int_{\Omega^s} W(\mathbf{x} - \mathbf{x}')d\mathbf{x}' = 1 \qquad (2.3)$$

which is the **unity condition** that any type of smoothing function must satisfy.

Since W is differentiable, we can now apply Gauss's divergence theorem to Eq. (2.1), leading to

$$\nabla U = \int_{\partial\Omega^s} U(\mathbf{x}')W(\mathbf{x}-\mathbf{x}')\vec{n}dS - \int_{\Omega^s} U(\mathbf{x}')\nabla W(\mathbf{x}-\mathbf{x}')d\mathbf{x}' \qquad (2.4)$$

where

- $\partial\Omega^s$ represents the boundary of domain Ω^s,
- $\vec{n}$ denotes the unit normal vector with components n_α on the boundary $\partial\Omega^s$ (see Figure 2.1).

Under these conditions, there can be many smoothing functions. However, we want the simplest possible ones. One such a W is the constant in Ω^s including its boundary $\partial\Omega^s$. The *unity condition* of W in Eq. (2.3) yields

$$\begin{cases} W(\mathbf{x}) = \dfrac{1}{V}, & \forall\mathbf{x} \text{ in } \Omega^s \text{ and on } \partial\Omega^s \\ W(\mathbf{x}) = 0, & \forall\mathbf{x} \notin \Omega^s \end{cases} \qquad (2.5)$$

where V is the volume of the smoothing domain Ω_s in the 3D case and the area in the 2D case. This form of smoothing function is called constant GSM.

Another form of smoothing function is piecewise linearly in the smoothing domain and vanishes on and beyond the boundary of the smoothing domain. This will be presented in Section 2.2.

2.1.2 *Gradient smoothing operator with constant smoothing function*

Using Eq. (2.5), the second term on the right-hand side of Eq. (2.4) vanishes. We thus have

$$\nabla U|_{GSM} = \frac{1}{V}\int_{\partial\Omega^s} U(\mathbf{x}')\vec{n}dS \qquad (2.6)$$

Note that the gradient given in Eq. (2.6) is a vector of partial derivatives of the function, $\partial U/\partial x_\alpha$. It is seen that a spatial derivative of the function $\partial U/\partial x_\alpha$ is now approximated by averaging the components of the function projected on the corresponding coordinate direction n_α on $\partial\Omega^s$. The direction component n_α keeps track of the contribution of the function in the direction, and the integration measures the rate of the change in function value on $\partial \Omega^s$. It is essentially a kind of finite difference, but now the curve integral ensures that the effects of the domain shape change are fully accounted for. This is the fundamental reason for the GSM approximation being orthogonal projection when measured in a weakened weak form [1]. It enables the GSM to work well with unstructured grids when discretization is implemented.

There is no specific rule for choosing the smoothing domains. However, if the smoothing domains are chosen in such a way that (1) their union covers the entire problem domain and (2) there is no gap or overlap between all these local smoothing domains, then the GSM approximation will be conservative, meaning that the influence from any neighbor of a certain point on itself is anti-symmetric to the influence from the point to this neighbor in derivative approximation. This is crucial for maintaining the conservation law when solving conservation equations or when conservation is a critical concern.

By successively applying the gradient smoothing to second-order derivatives, the Laplace operator at the location $\mathbf{x}$ can be approximated as

$$\nabla \cdot (\nabla U)|_{GSM} = \frac{1}{V} \int_{\partial \Omega^s} \nabla U(\mathbf{x}') \cdot \vec{n} dS \qquad (2.7)$$

In GSM implementations, we often need the second derivatives of the function at a point where the PDEs are discretized. This is done using the gradient of the function on the smoothing domain boundary used in Eq. (2.7). Therefore, the smoothing domain used in Eq. (2.7) is usually different from that used in (2.6) to compute the gradient of the function.

2.1.3 *Construction of gradient smoothing domains*

We now discuss the means to construct the *gradient smoothing domain* or GSD.

Assume that the problem domain is represented by a set of nodes or particles. These nodes can be used to construct a mesh using, for example, the Delaunay triangulation techniques, which result in unstructured triangular cells and edges. The simplest way to construct the smoothing domain is to

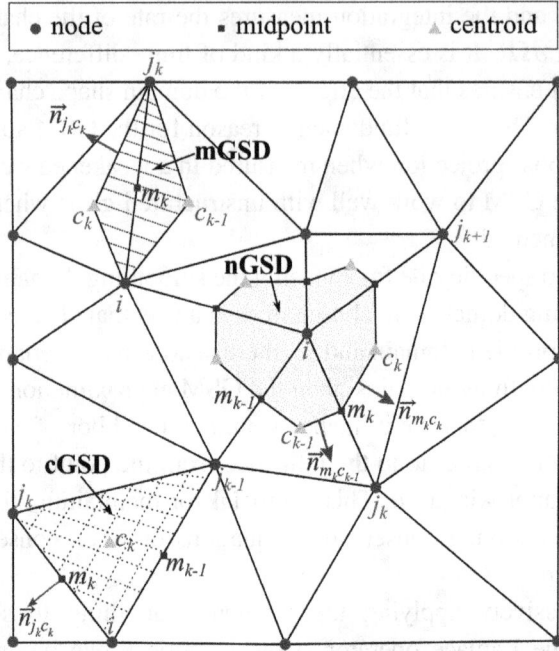

Figure 2.2. Illustration of node-associated GSD (nGSD), midpoint-associated GSD (mGSD) and centroid-associated GSD (cGSD). Node: i or j. Midpoint: m_k. Centroid: c_k.

make use of the geometric characteristic points: the nodes, the midpoints of the edges, and the centroids of the triangular cells (see Figure 2.2).

Based on the location of the point of interest, there are three commonly used types of GSDs:

- node-associated GSD (nGSD),
- centroid-associated GSD (cGSD),
- midpoint-associated GSD (mGSD), as illustrated in Figure 2.2.

The *first* type of smoothing domain is the node-associated and is hence denoted as nGSD. It is used for approximating the derivatives at a node or particle of interest and is formed by sequentially connecting the centroids of triangles with the midpoints of the edges connected to the node.

The *second* is formed by a primitive cell called cGSD. It is used for approximating the derivatives at the centroid of the cell. This is similar to the cell-centered finite volume method (FVM).

The *third* is midpoint-associated or mGSD. It is used for the calculation of gradients at the midpoint of a cell-edge of interest. The mGSD is formed by connecting the end-nodes of cell-edge with the centroids on both sides of the cell-edge, as shown in Figure 2.2.

With Eqs. (2.6) and (2.7), the spatial derivatives at any point of interest can be approximated based on the corresponding types of GSDs described above. Different schemes are devised for this purpose, which will be elucidated further in the following section.

2.1.4 *Spatial approximation schemes*

To approximate integrals along the boundaries of various types of GSDs, two quadrature rules have been proposed [8]: *one-point* quadrature (rectangular rule) and *two-point* quadrature (trapezoidal rule).

In the *one-point* quadrature scheme, the integration along the smoothing domain-edge is done using the rectangular rule, and the integrand is evaluated only at the midpoint of a cell-edge. In the *two-point* quadrature scheme, the trapezoidal rule is used, and the integrand is evaluated at the two end-points of a smoothing domain-edge.

Generally, approximating derivatives of field variables on nodes needs the value or gradient of field variables on midpoints and/or centroids. The value and gradient of field variables in those locations can be evaluated by either linear interpolation (LI) or gradient smoothing (GS) based on the corresponding type of GSD, as described in Section 2.1.3.

LI at the midpoints has the form of

$$U_{m_k} \approx \frac{U_i + U_{j_k}}{2}, \quad \nabla U_{m_k} \approx \frac{\nabla U_i + \nabla U_{j_k}}{2} \tag{2.8}$$

and at the centroids

$$U_{c_k} \approx \frac{U_i + U_{j_k} + U_{j_k+1}}{3}, \quad \nabla U_{c_k} \approx \frac{\nabla U_i + \nabla U_{j_k} + \nabla U_{j_k+1}}{3} \tag{2.9}$$

2.1.4.1 *One-point quadrature schemes*

The one-point quadrature assumes uniform value and gradient of field variables on each boundary division $\overline{c_{k-1}m_kc_k}$, i.e.,

$$U_{c_k} = U_{c_k-1} = U_{m_k} \tag{2.10}$$

$$\nabla U_{c_k} = \nabla U_{c_k-1} = \nabla U_{m_k} \tag{2.11}$$

where U_m and U_c denote the field variable U at midpoints of cell-edges and centroids of triangular cells, respectively. Subscript k indicates the kth supporting nodes of the target node.

As a consequence, Eqs. (2.6) and (2.7) can be simplified as

$$
\begin{cases}
\dfrac{\partial U_i}{\partial x} \approx \dfrac{1}{V_i} \sum\limits_{k=1}^{n_i} U_{m_k}(\Delta S_x)_{ijk} \\[2ex]
\dfrac{\partial U_i}{\partial y} \approx \dfrac{1}{V_i} \sum\limits_{k=1}^{n_i} U_{m_k}(\Delta S_y)_{ijk}
\end{cases}
\tag{2.12}
$$

and

$$\nabla \cdot (\nabla U_i) \approx \frac{1}{V_i} \sum_{k=1}^{n_i} \left[\frac{\partial U_{m_k}}{\partial x}(\Delta S_x)_{ijk} + \frac{\partial U_{m_k}}{\partial y}(\Delta S_y)_{ijk} \right] \tag{2.13}$$

where n_i denotes the total number of supporting nodes within the stencil of node i:

$$
\begin{cases}
(\Delta S_x)_{ijk} = (\Delta S_x)_{ijk}^{(L)} + (\Delta S_x)_{ijk}^{(R)} \\[1.5ex]
(\Delta S_y)_{ijk} = (\Delta S_y)_{ijk}^{(L)} + (\Delta S_y)_{ijk}^{(R)}
\end{cases}
\tag{2.14}
$$

The face vectors of the two respective domain-edges are calculated as

$$(\Delta S_x)_{ijk}^{(L)} = \Delta S_{ijk}^{(L)}(n_x)_{ijk}^{(L)} \tag{2.15}$$

$$(\Delta S_x)_{ijk}^{(R)} = \Delta S_{ijk}^{(R)}(n_x)_{ijk}^{(R)} \tag{2.16}$$

$$(\Delta S_y)_{ijk}^{(L)} = \Delta S_{ijk}^{(L)}(n_y)_{ijk}^{(L)} \tag{2.17}$$

$$(\Delta S_y)_{ijk}^{(R)} = \Delta S_{ijk}^{(R)}(n_y)_{ijk}^{(R)} \tag{2.18}$$

Superscripts (L) and (R) are pointers to the two domain-edges associated with the cell-edge of interest, ij_k. ΔS_x and ΔS_y are the two components of a domain-edge vector. n_x and n_y represent the two components of the unit normal vector $\vec{n}$ of the domain-edge $\partial \Omega_i$.

Taking nGSD as an example, the components of normal vectors occurring in Eqs. (2.15)–(2.18) are calculated as

$$(n_x)^{(L)}_{ijk} = (n_x)_{m_k c_k} = \frac{y_{c_k} - y_{m_k}}{r_{m_k c_k}} \tag{2.19}$$

$$(n_x)^{(R)}_{ijk} = (n_x)_{c_{k-1} m_k} = \frac{y_{m_k} - y_{c_{k-1}}}{r_{c_{k-1} m_k}} \tag{2.20}$$

$$(n_y)^{(L)}_{ijk} = (n_y)_{m_k c_k} = \frac{x_{c_k} - x_{m_k}}{r_{m_k c_k}} \tag{2.21}$$

$$(n_y)^{(R)}_{ijk} = (n_y)_{c_{k-1} m_k} = \frac{x_{m_k} - x_{c_{k-1}}}{r_{c_{k-1} m_k}} \tag{2.22}$$

where $r_{m_k c_k}$ and $r_{c_{k-1} m_k}$ represent the lengths of $m_k c_k$ and $c_{k-1} m_k$, respectively.

First-order derivative at nodes: With the two-point quadrature rule, the first-order derivatives represented by Eq. (2.6) of the field variable U at node i are given by

$$\begin{cases} \dfrac{\partial U_i}{\partial x} \approx \dfrac{1}{V_i} \sum\limits_{k=1}^{n_i} \left[\dfrac{U_{m_k} + U_{c_k}}{2} (\Delta S_x)^{(L)}_{ijk} + \dfrac{U_{m_k} + U_{c_{k-1}}}{2} (\Delta S_x)^{(R)}_{ijk} \right] \\[4mm] \dfrac{\partial U_i}{\partial y} \approx \dfrac{1}{V_i} \sum\limits_{k=1}^{n_i} \left[\dfrac{U_{m_k} + U_{c_k}}{2} (\Delta S_y)^{(L)}_{ijk} + \dfrac{U_{m_k} + U_{c_{k-1}}}{2} (\Delta S_y)^{(R)}_{ijk} \right] \end{cases} \tag{2.23}$$

where the face vectors of the two respective domain-edges have the forms as shown in Eqs. (2.15)–(2.18).

The field variable U at non-storage locations, i.e., at midpoints and centroids, are computed by simple interpolation represented by Eqs. (2.8) and (2.9).

First-order derivative at midpoints: Analogous to the discretization at nodes described above, the gradient at midpoint ∇U_{m_k} can also be approximated with the gradient smoothing technique in Eq. (2.6) but based on the related mGSD (see Figure 2.2).

Specifically,

$$\frac{\partial U_{m_k}}{\partial x} \approx \frac{1}{V_{m_k}} \left[\frac{U_i + U_{c_{k-1}}}{2} (\Delta S_m^x)_{ic_{k-1}} + \frac{U_{c_{k-1}} + U_{j_k}}{2} (\Delta S_m^x)_{c_{k-1}j_k} \right.$$
$$\left. + \frac{U_{j_k} + U_{c_k}}{2} (\Delta S_m^x)_{j_k c_k} + \frac{U_{c_k} + U_i}{2} (\Delta S_m^x)_{c_k i} \right] \qquad (2.24)$$

$$\frac{\partial U_{m_k}}{\partial y} \approx \frac{1}{V_{m_k}} \left[\frac{U_i + U_{c_{k-1}}}{2} (\Delta S_m^y)_{ic_{k-1}} + \frac{U_{c_{k-1}} + U_{j_k}}{2} (\Delta S_m^y)_{c_{k-1}j_k} \right.$$
$$\times \frac{U_{j_k} + U_{c_k}}{2} (\Delta S_m^y)_{j_k c_k} + \frac{U_{c_k} + U_i}{2} (\Delta S_m^y)_{c_k i} \right] \qquad (2.25)$$

where ΔS_m^x and ΔS_m^y denote the components of a respective face vector for an mGSD of interest and V_m stands for the area or volume of the mGSD. These face vectors can be calculated similar to those for nGSD.

First-order derivative at centroids: Similarly, the gradients at centroids ∇U_{m_k} can be evaluated with gradient smoothing technique over cGSD by

$$\frac{\partial U_{c_k}}{\partial x} \approx \frac{1}{V_{c_k}} \left[\frac{U_i + U_{j_k}}{2} (\Delta S_c^x)_{ij_k} + \frac{U_{j_k} + U_{j_{k+1}}}{2} (\Delta S_c^x)_{j_k j_{k+1}} \right.$$
$$\left. + \frac{U_{j_{k+1}} + U_i}{2} (\Delta S_c^x)_{j_{k+1}i} \right] \qquad (2.26)$$

$$\frac{\partial U_{c_k}}{\partial y} \approx \frac{1}{V_{c_k}} \left[\frac{U_i + U_{j_k}}{2} (\Delta S_c^y)_{ij_k} + \frac{U_{j_k} + U_{j_{k+1}}}{2} (\Delta S_c^y)_{j_k j_{k+1}} \right.$$
$$\left. + \frac{U_{j_{k+1}} + U_i}{2} (\Delta S_c^y)_{j_{k+1}i} \right] \qquad (2.27)$$

with ΔS_c^x and ΔS_c^y being the two components of a respective face vector for a cGSD of interest. V_c is the volume or area of the cGSD.

Second-order derivatives at nodes:

The calculation of second-order derivatives at nodes of interest is necessary for solving the N−S governing equations involving a viscous term. By using the gradient smoothing technique, they can be computed in the following form:

$$
\frac{\partial^2 U_i}{\partial x^2} + \frac{\partial^2 U_i}{\partial y^2} \approx \frac{1}{V_i} \sum_{k=1}^{n_i} \frac{1}{2} \left\{ \left[\frac{\partial U_{m_k}}{\partial x} + \frac{\partial U_{c_k}}{\partial x} \right] (\Delta S_x)_{ij_k}^{(L)} \right.
$$

$$
+ \left[\frac{\partial U_{m_k}}{\partial y} + \frac{\partial U_{c_k}}{\partial y} \right] (\Delta S_y)_{ij_k}^{(L)}
$$

$$
+ \left[\frac{\partial U_{m_k}}{\partial x} + \frac{\partial U_{c_{k-1}}}{\partial x} \right] (\Delta S_x)_{ij_k}^{(R)}
$$

$$
\left. + \left[\frac{\partial U_{m_k}}{\partial y} + \frac{\partial U_{c_{k-1}}}{\partial y} \right] (\Delta S_y)_{ij_k}^{(R)} \right\} \tag{2.28}
$$

All the first-order derivatives at midpoints and centroids are approximated as presented in Eqs. (2.24)–(2.27).

As shown in Eqs. (2.12) and (2.13), in one-point quadrature schemes, only values for the field variable and its gradients at midpoints are needed. Thus, the vectors for a pair of domain-edges connected with the cell-edge ij_k can be lumped together, which in return reduces the storage space for geometrical parameters. Apparently, the one-point quadrature schemes are simpler and much more cost-effective in terms of both flops and storage.

In contrast, the two-point quadrature imposes extra requirements in computation and storage for values of variables at centroids and domain-edge vectors for cGSD. When mGSDs are used for the prediction of gradients at midpoints, such demands become even higher. However, theoretically, schemes based on two-point quadrature may give more accurate results. This will be verified and discussed later using numerical examples.

2.1.5 *Effect of spatial approximation on solution accuracy*

To investigate the effect of these spatial approximation strategies on solution accuracy, we proposed a total of eight discretizing schemes for second-order derivative approximation at nodes with Eq. (2.28), as listed in Table 2.1, by combining different quadrature rules with different types of GSDs.

Table 2.1.   Some approximation schemes of the second-order spatial derivatives.

Schemes	Quadrature	Type of GSD	Approximation of gradients at midpoints	Approximation of gradients at centroids	Directional correction
I	One-point	nGSD	LI	Not required	No
II	One-point	nGSD	LI	Not required	Yes
III	Two-point	nGSD	LI	LI	No
IV	Two-point	nGSD	LI	LI	Yes
V	Two-point	nGSD, cGSD	LI	GS	No
VI	Two-point	nGSD, cGSD	LI	GS	Yes
VII	One-point	nGSD, mGSD	GS	Not required	No
VIII	Two-point	nGSD, mGSD, cGSD	GS	GS	No

Note: LI = Linear Interpolation, GS = Gradient Smoothing.

The gradients at midpoint and centroid are required to approximate the second-order derivative at node with Eq. (2.28). The gradient at the midpoint of a cell-edge can be calculated in two ways: either by simple interpolation using the gradients at both end-nodes of the cell-edge (I–VI) or by gradient smoothing technique using Eq. (2.6) based on mGSD (VII and VIII). Similarly, the gradients at a centroid can be obtained either by simple interpolation using the gradients at the nodes of its corresponding cGSD (III and IV) or by gradient smoothing over the corresponding cGSD (V, VI and VIII).

In Table 2.1, the linear interpolation LI takes the form in Eqs. (2.8) and (2.9). Note that when one-point quadrature scheme is used, there is no need to approximate the gradients at centroids, since the integrands in Eqs. (2.6) and (2.7) are evaluated only at the midpoints of cell-edges.

We found that as the gradients at midpoint are estimated using Eqs. (2.24) and (2.25), decoupling solutions (checkerboard problem) are attained [4]. To circumvent such a problem, as proposed in [9], the approximated gradients at midpoint are remedied with the **directional correction** technique:

$$\nabla \tilde{U}_{m_k} = \nabla U_{m_k} - \left[\nabla U_{m_k} \cdot \vec{l}_{ij_k} - \left(\frac{\partial U}{\partial l} \right)_{ij_k} \right] \tag{2.29}$$

where

$$\left(\frac{\partial U}{\partial l} \right)_{ij_k} \approx \frac{U_{j_k} - U_i}{\Delta l_{ij_k}} \tag{2.30}$$

$$\vec{l}_{ij_k} = \mathbf{x}_{j_k} - \mathbf{x}_i \tag{2.31}$$

$$\Delta l_{ij_k} = |\mathbf{x}_{j_k} - \mathbf{x}_i| \tag{2.32}$$

Here, $\mathbf{x}_i$ and $\mathbf{x}_{j_k}$ denote the spatial locations of node i and j_k, respectively.

This directional correction technique is adopted in schemes II, IV and VI because the GSM is not used for values at the middle point of the edges. Details about the role of directional correction will be addressed in the following stencil analysis. For schemes that used only GSM for approximation (VII and VIII), directional correction is not needed.

2.1.6 *Stencil analysis*

A careful investigation of the stencils of supporting nodes for the proposed schemes is carried out. This study helps to select the most suitable schemes. In order to obtain explicit results, square and equilateral triangle cells are used in the study.

2.1.6.1 *Basic principles for stencil assessment*

In the stencil analyses, the following five basic rules are considered to assess these discretization schemes [8]:

(a) *consistency* at each interface of two adjacent gradient smoothing domains,
(b) *positivity* of coefficients of influence,
(c) *sum* of coefficients of influence,
(d) *negative*-slope linearization of the source term,
(e) *compactness* of the stencil.

The first four rules were initially summarized in [10] considering physically realistic behavior and overall balance. The last one is for the sparseness of the resulting system equations.

Rule (a) requires to use the same expression of approximation on the interface of two adjacent GSDs.

Rule (b) requires that the coefficient for the node of interest and the coefficients of influence must be positive, once the discretization equation is written in the form of

$$a_{ii}U_i + \sum_{k=1}^{n_i} a_{ij_k}U_{j_k} = b_i$$

Rule (c) requires

$$a_{ii} = -\sum_{k=1}^{n_i} a_{ij_k}$$

Rule (d) relates to the treatment of the source terms. As addressed by Patankar [10], it is essential to keep the slope of linearization to be negative, since a positive slope can lead to computational instabilities and physically unrealistic solutions.

It is also necessary for a good discretization stencil to satisfy Rule (e) for efficiency, as commented by Barth [11]. The very first layer of nodes surrounding the node of interest should be included in the discretization stencil. Moreover, as the stencil becomes larger, not only does the computational cost increase but eventually the accuracy also decreases as less valid data from further away are brought into approximation.

Besides, a few lemmas have been proposed [11] to address the necessity of positivity of coefficients to satisfy a discrete maximum principle for numerical schemes suitable for non-oscillatory discontinuity (e.g., shock). At steady state, the non-negativity of the coefficients becomes sufficient to satisfy a discrete maximum principle for stable results. His statements reiterate the importance of Rule (b) [10].

In GSM, when the gradient smoothing technique is applied to the GSDs, Rule (a) is automatically satisfied, meaning that the local conservation of quantities is ensured. It also ensures global conservation once proper boundary conditions are used. In the following sections, Rules (b), (c) and (e) are assessed for stencils for schemes given in Table 2.1.

2.1.6.2 *Stencils for approximated gradients*

The stencils for gradient approximation using the eight types of discretization schemes, as listed in Table 2.1, are derived. The coefficients of influence on cells with square and equilateral triangle shapes of unit length are shown in Figures 2.3 and 2.4.

Square cells: We find that the three one-point quadrature schemes (I, II and VII) give the same stencil when square cells are used, as shown in Figure 2.3(a). This stencil is also identical to that obtained by using

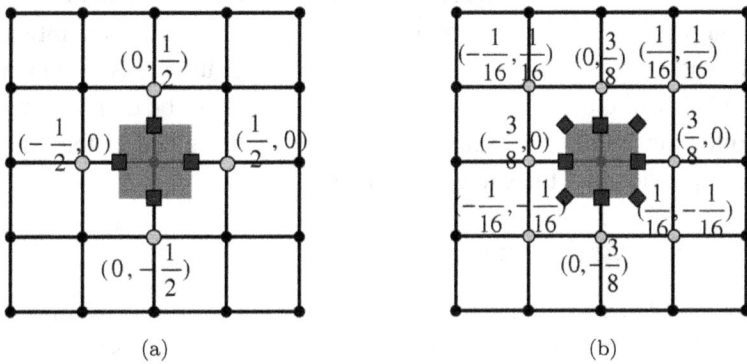

(a)   (b)

Figure 2.3. Stencils for approximated gradients based on square cells: (a) I, II and VII; (b) III, IV, V, VI, and VIII (Reprinted with permission from [8]).

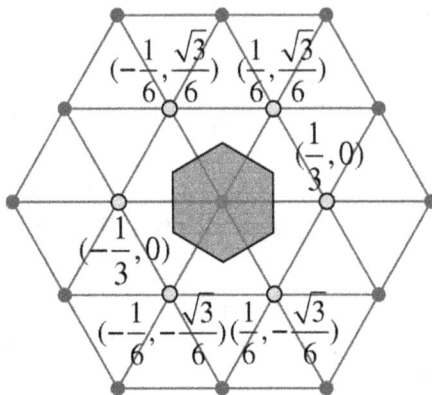

Figure 2.4. Stencil for approximated gradients based on the cells in equilateral triangular shape (identical for all the eight schemes) (Reprinted with permission from [8]).

the two-point central-difference scheme in the FDM because the grids are regular. Moreover, it has been noted that the stencil for all two-point quadrature schemes is the same, as shown in Figure 2.3(b). This stencil is also identical to the one obtained by using the six-point central-difference scheme in the FDM. Such findings confirm that when square cells are used, the GSM is identical to the FDM. The GSM, however, works for cells in irregular shapes.

Cells in equilateral triangle shape: It is interesting that the stencil based on equilateral triangular cells is found to be the same for all the GSM schemes, as shown in Figure 2.4. This stencil is identical to that of the interpolation method using six surrounding nodes [1]. However, the interpolation method may not work well for irregular triangular cells as addressed by Liu [1], while the GSM still performs well, as will be demonstrated in the following numerical examples. This can be attributed to the inherent stability provided by the smoothing operation.

2.1.6.3 *Stencils for approximated Laplace operator*

Square cells: The stencils for the approximated Laplace operator with GSM schemes based on square cells are derived and presented in Figure 2.5. Three of the schemes, I, III, and V, as shown in Figure 2.5(a), (c) and (e), result in wide stencils with *unfavorable* weighting coefficients (zero and negative) on square cells. Such stencils may lead to unexpected decoupling solutions as confirmed in the current analysis and numerical examples. These *unfavorable* schemes cannot effectively dampen high-frequency numerical errors as mentioned by Moinier [12] and may perform even worse in the boundary layer region, where the viscous effect becomes dominant.

It is found that such unfavorable stencils result from the simple interpolation used to approximate the gradients at the midpoints of cell-edges in the three schemes. In contrast, schemes II and VI, which incorporate directional correction to the approximated gradients at midpoints, yield relatively compact stencils with favorable coefficients, as depicted in Figures 2.5(b) and (f). Scheme II is a five-point stencil while Scheme VI corresponds to a nine-node compact stencil, which is the same as those for the central-difference scheme in the FDM. However, as shown in

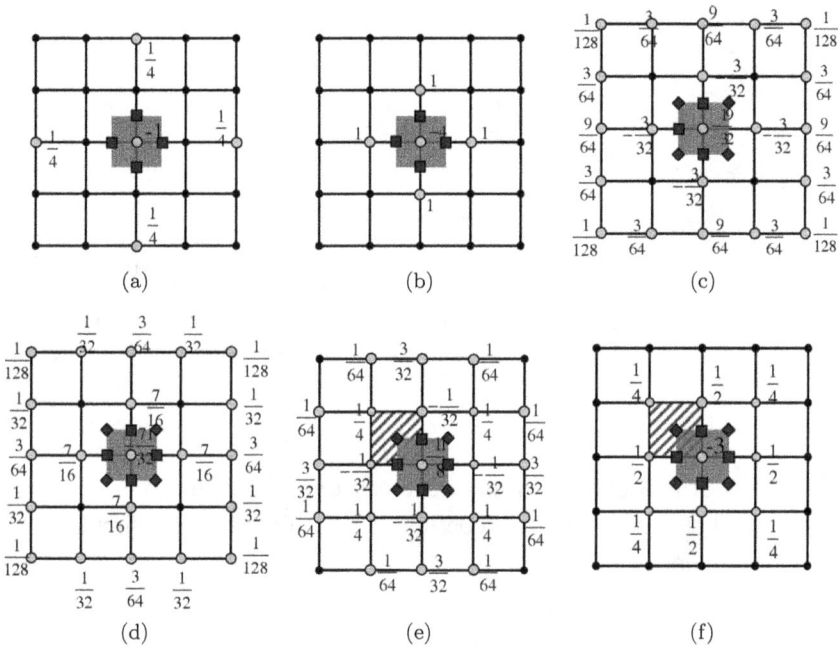

Figure 2.5. Stencils for approximated Laplace operator on square cells: (a) I; (b) II and VII; (c) III; (d) IV; (e) V; (f) VI and VIII (Reprinted with permission from [8]).

Figure 2.5(d), even with directional correction, an unfavorable stencil may still occur in Scheme IV. Thus, Schemes II and VI are labeled as favorable, while Scheme IV is *unfavorable*.

Furthermore, compact and favorable stencils are obtained using Schemes VII and VIII, where the gradients at midpoints of cell-edges are approximated by applying gradient smoothing operation to mGSD, as seen in Figures 2.5(b) and (f). These two schemes do not require any direction correction, implying that they are naturally consistent and stable. From the perspectives of the consistency in the approximation of derivatives at different locations and the stability feature of the gradient smoothing technique, Schemes VII and VIII are more robust and preferable in general settings than Schemes II and VI.

Cells in equilateral triangle shape: Similar analyses are conducted by using equilateral triangular cells and the resulting stencils are shown in

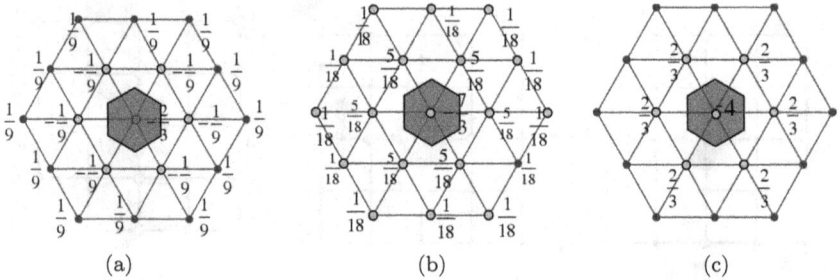

(a)   (b)   (c)

Figure 2.6. Stencils for approximated Laplace operator on the cells in equilateral triangular shape: (a) I and III; (b) IV and V; (c) II, VI, VII and VIII (Reprinted with permission from [1]).

Figure 2.6. It is found again that Schemes I and III result in an *unfavorable* stencil due to the negative coefficients of influence at the first layer of neighboring nodes, as seen in Figure 2.6(a), violating Rule (b). On the other hand, Schemes IV and V, and Schemes II, VI, VII and VIII produce two sets of stencils with favorable coefficients, as shown in Figure 2.6(b) and (c), respectively. Schemes IV and V yield a two-layer surrounding points stencil around the node of interest, while Schemes II, VI, VII and VIII use only the first layer of neighboring nodes. According to Rule (e), the latter set of schemes is more *favorable* because of their relatively more compact stencils.

It is important to note that all schemes studied here satisfy Rule (c). Based on the stencil analyses, four favorable GSM schemes, namely, Schemes II, VI, VII, and VIII, are selected for further examination, since these schemes consistently produce compact stencils with favorable coefficients for both square and equilateral triangular cells. Schemes II and VI are partial GSM schemes that require directional correction, while Schemes VII and VIII are full GSM schemes that do not require directional correction and are expected to perform well for irregular cells.

2.1.6.4 *Truncation errors*

We next conduct analyses on truncation errors for the four recommended schemes for approximating first- and second-order derivatives based on square and equilateral triangular cells. The results are summarized in

Table 2.2. Truncation errors in the approximation of first derivatives in GSM.

Schemes	Shapes of cells	Truncation error
II and VII	Square	$\begin{cases} O_x(h^2) = -\frac{h^2}{6}\frac{\partial^3 U_{ij}}{\partial x^3} + O(h^3) \\ O_y(h^2) = -\frac{h^2}{6}\frac{\partial^3 U_{ij}}{\partial y^3} + O(h^3) \end{cases}$
VI and VIII	Square	$\begin{cases} O_x(h^2) = -h^2\left(\frac{5}{24}\frac{\partial^3 U_{ij}}{\partial x^3} + \frac{1}{2}\frac{\partial^3 U_{ij}}{\partial xy^2}\right) + O(h^3) \\ O_y(h^2) = -h^2\left(\frac{5}{24}\frac{\partial^3 U_{ij}}{\partial y^3} + \frac{1}{2}\frac{\partial^3 U_{ij}}{\partial x^2 y}\right) + O(h^3) \end{cases}$
II, VI, VII and VIII	Equilateral triangle	$\begin{cases} O_x(h^2) = -h^2\left(\frac{1}{24}\frac{\partial^3 U_{ij}}{\partial x^3} + \frac{1}{8}\frac{\partial^3 U_{ij}}{\partial xy^2}\right) + O(h^3) \\ O_y(h^2) = -h^2\left(\frac{1}{24}\frac{\partial^3 U_{ij}}{\partial y^3} + \frac{1}{8}\frac{\partial^3 U_{ij}}{\partial x^2 y}\right) + O(h^3) \end{cases}$

Table 2.3. Truncation errors in the GSM approximation of the Laplace operator.

Schemes	Shapes of cells	Truncation error
II and VII	Square	$O(h^2) = -\frac{h^2}{12}\left(\frac{\partial^4 U_{ij}}{\partial x^4} + \frac{\partial^4 U_{ij}}{\partial y^4}\right) + O(h^3)$
VI and VIII	Square	$O(h^2) = -\frac{h^2}{12}\left(\frac{\partial^4 U_{ij}}{\partial x^4} + 3\frac{\partial^4 U_{ij}}{\partial x^2 y^2} + \frac{\partial^4 U_{ij}}{\partial y^4}\right) + O(h^3)$
II, VI, VII and VIII	Equilateral triangle	$O(h^2) = -\frac{h^2}{16}\left(\frac{\partial^4 U_{ij}}{\partial x^4} + 2\frac{\partial^4 U_{ij}}{\partial x^2 y^2} + \frac{\partial^4 U_{ij}}{\partial y^4}\right) + O(h^3)$

Tables 2.2 and 2.3 [13]. All of these schemes are of second-order accuracy. Scheme VII has the same truncation errors as Scheme II, while Schemes VIII and VI have identical truncation errors. These theoretical findings will be further validated when these schemes are used to solve Poisson equations in Chapter 3.

2.2 Linearly-weighted gradient smoothing

In Section 2.1, a *constant* smoothing function is adopted in the gradient smoothing operator. The use of the *constant* smoothing function leads to a simplified form of integrand along the domain-edges for approximating spatial derivatives.

In this section, we present a *linearly-weighted* gradient smoothing method (LW-GSM) using a linear smoothing function.

2.2.1 Construction of piecewise linear smoothing functions

Recall that the gradient at a node of interest can be represented by Eq. (2.4) in Section 2.1.1. This equation is also applicable to *linearly-weighted* form [14]. Consider a one-dimensional (1D) smoothing domain of $\Omega_{\mathbf{x}}^s = [x_L, x_R]$. The linear smoothing function over the smoothing domain of $\Omega_{\mathbf{x}}^s$ containing x can be written as

$$W(x-x') = \begin{cases} \dfrac{-hx_L}{x-x_L} + \dfrac{h}{x-x_L}x' & \forall x' \in [x_L, x] \\ \dfrac{hx_R}{x_R-x} + \dfrac{-h}{x_R-x}x' & \forall x' \in [x, x_R] \\ 0 & \text{else} \end{cases} \tag{2.33}$$

where h is the height of the function given by

$$h = \frac{2}{x_R - x_L} \tag{2.34}$$

A 2D smoothing domain $\Omega_{\mathbf{x}}^s$ for a point of interest $\mathbf{x}$ can be expressed as

$$W(\mathbf{x}-\mathbf{x}') = \begin{cases} a_0 + \mathbf{a}_1 \cdot (\mathbf{x}-\mathbf{x}'), & \mathbf{x}' \in \Omega_{\mathbf{x}}^s \\ 0, & \mathbf{x}' \notin \Omega_{\mathbf{x}}^s \end{cases} \tag{2.35}$$

where a_0 is a constant and vector $\mathbf{a}_1$ is a vector containing two constants. All these three constants depend on the geometry of individual sub-triangle within the smoothing domain. Figure 2.7 depicts the spatial distribution of piecewise linear smoothing functions over the three types of smoothing domains for approximating gradients at different locations.

To approximate spatial derivatives effectively, three constraints are imposed for the proposed linear smoothing function:

(1) Similar to the piecewise *constant* smoothing function in the standard GSM, the consistency of gradient approximation requires the piecewise *linear* smoothing function in the LW-GSM to satisfy the unity condition:

$$\sum_{k=1}^{2n_i} \int_{\Omega_i^{(k)}} W(\mathbf{x}-\mathbf{x}') d\mathbf{x}' = 1 \tag{2.36}$$

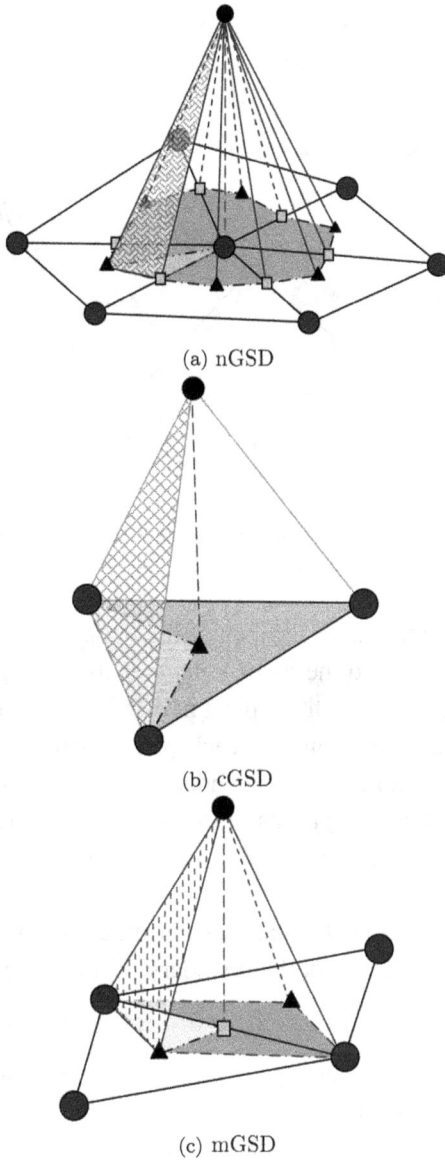

(a) nGSD

(b) cGSD

(c) mGSD

Figure 2.7. Linearly-weighted smoothing functions for different types of gradient smoothing domains subjected to piecewise linear smoothing functions (Reprinted with permission from [14]).

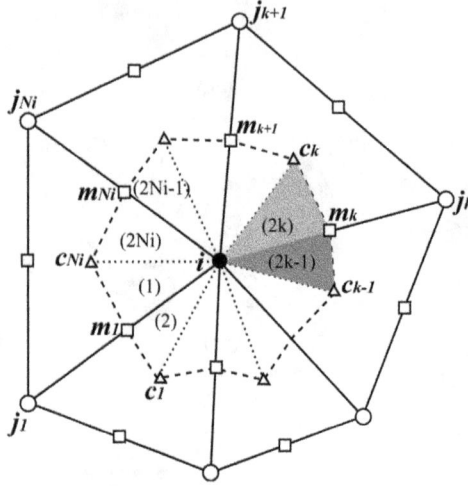

Figure 2.8. Adopted notations and sub-triangulation in the nGSD for node i of interest (Reprinted with permission from [14]).

where the total number of the supporting nodes for the node of interest, i, is denoted by n_i and the index k denotes its kth supporting node. As illustrated in Figure 2.8, the typical nGSD for an interior node consists of a total of $2n_i$ sub-triangles, each pair of which is connected with a common grid edge. For a boundary node as shown in Figure 2.9, its smoothing domain consists of $2n_i - 2$ sub-triangles. It is noted that the sub-triangles are considered in pairs in the derivation of LW-GSM formulations.

(2) To simplify the gradient smoothing formulation, the linear smoothing functions are set to zero at the boundary of the domain, i.e.,

$$W(\mathbf{x} - \mathbf{x}_i) = 0, \quad \mathbf{x} \in \partial\Omega_i^s \tag{2.37}$$

With the above equation, the first integral term over $\partial\Omega$ in Eq. (2.4) vanishes automatically.

(3) The piecewise linear smoothing function is assumed to be continuous within the nGSD of interest. Together with the zero value for regions beyond the nGSD as shown in Eq. (2.35) and the second constraint represented by Eq. (2.37), the proposed piecewise linear smoothing function is continuous across the whole field. Besides, as shown in Figure 2.7, the third constraint also implies a constant value of the

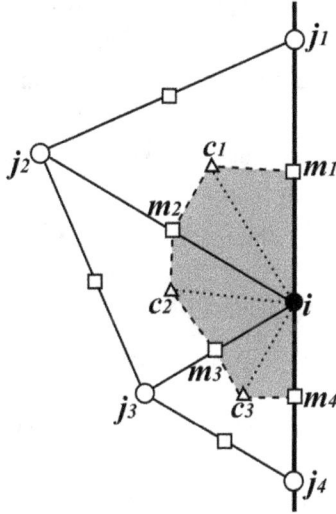

Figure 2.9. Smoothing domain for a node on the symmetrical boundary (Reprinted with permission from [14]).

smoothing function at the node of interest, i.e.,

$$W(\mathbf{x} - \mathbf{x}_i) = a_0, \quad \text{as } \mathbf{x} = \mathbf{x}_i \qquad (2.38)$$

Hence, it is apparent that across the nGSD for the node i, the piecewise linear smoothing function, W, is continuous, but its derivatives, ∇W, are discontinuous.

2.2.2 *Linear-weighted gradient smoothing operator*

Considering the second constraint of the linear-weighted smoothing function represented by Eq. (2.37), Eq. (2.4) can be simplified as

$$\nabla U \approx -\int_{\Omega^s} U(\mathbf{x}')\nabla W(\mathbf{x} - \mathbf{x}')d\mathbf{x}' \qquad (2.39)$$

Analogously, the Laplace operator at node i can be approximated in the simplified form of

$$\nabla \cdot \nabla U \approx -\int_{\Omega^s} \nabla U(\mathbf{x}') \cdot \nabla W(\mathbf{x} - \mathbf{x}')d\mathbf{x}' \qquad (2.40)$$

As shown in Eqs. (2.39) and (2.40), the integrals over the smoothing domain are engaged in the approximations once the piecewise linear functions are adopted as the smoothing functions. In Eq. (2.40), the integrand is the product of the gradients of the smoothing functions and the spatial derivative of concern in reduced order. Thus, an additional approximation of the gradients of the smoothing function is demanded in the LW-GSM. It is apparent that the approximation formulations based on the piecewise linear smoothing functions take on totally different forms from those in the standard GSM that is based on constant smoothing functions.

2.2.2.1   *Determination of smoothing functions*

In the 2D case, the piecewise linear smoothing function can be rewritten as

$$W(\mathbf{x} - \mathbf{x}_i) = \mathbf{a}_i + \mathbf{b}_i(x - x_i) + \mathbf{c}_i(y - y_i) \tag{2.41}$$

with $\mathbf{a}_i$, $\mathbf{b}_i$, $\mathbf{c}_i$ being the coefficients matrices owning the form of

$$\mathbf{a}_i = \begin{bmatrix} a_{i,1} \\ a_{i,2} \\ \vdots \\ a_{i,2n_i-1} \\ a_{i,2n_i} \end{bmatrix}, \quad \mathbf{b}_i = \begin{bmatrix} b_{i,1} \\ b_{i,2} \\ \vdots \\ b_{i,2n_i-1} \\ b_{i,2n_i} \end{bmatrix}, \quad \mathbf{c}_i = \begin{bmatrix} c_{i,1} \\ c_{i,2} \\ \vdots \\ c_{i,2n_i-1} \\ c_{i,2n_i} \end{bmatrix} \tag{2.42}$$

The gradient of the smoothing function can be obtained easily by differentiating Eq. (2.41), yielding

$$\begin{cases} \dfrac{\partial W(\mathbf{x} - \mathbf{x}_i)}{\partial x} = \mathbf{b}_i \\[2mm] \dfrac{\partial W(\mathbf{x} - \mathbf{x}_i)}{\partial y} = \mathbf{c}_i \end{cases} \tag{2.43}$$

Let us focus on the approximation of spatial derivatives at an inner node, as shown in Figure 2.8. Equation (2.37) implies that for each sub-triangle, e.g., $\Delta im_k c_k$, following relations hold:

$$\begin{cases} W_{m_k} = W(\mathbf{x}_{m_k} - \mathbf{x}_i) = 0 \\ W_{c_k} = W(\mathbf{x}_{c_k} - \mathbf{x}_i) = 0 \end{cases} \tag{2.44}$$

With the help of Eqs. (2.36) and (2.38), it is readily obtained that

$$a_{i,2k} = a_{i,2k-1} = W(\mathbf{x}_i - \mathbf{x}_i) = \frac{3}{V_i} \qquad (2.45)$$

where V_i is the total area of the smoothing domain for node i.

The components of coefficient matrices $\mathbf{b}_i$ and $\mathbf{c}_i$ for each sub-triangle can then be obtained from the solutions to Eq. (2.44):

$$b_{i,2k} = -a_{i,2k}\frac{(y_{c_k} - y_{m_k})}{\chi} \qquad (2.46)$$

$$c_{i,2k} = a_{i,2k}\frac{(x_{c_k} - x_{m_k})}{\chi} \qquad (2.47)$$

with

$$\chi = (x_{m_k} - x_i)(y_{c_k} - y_i) - (x_{c_k} - x_i)(y_{m_k} - y_i) \qquad (2.48)$$

It should be noted that χ relates to the area of the sub-triangle Δim_kc_k, i.e.,

$$A_{i,2k} = \frac{|\chi|}{2} \qquad (2.49)$$

It is noted that if the constitutive nodes $(i \to m_k \to c_k)$ for the sub-triangle Δim_kc_k obey the right-hand rule ordering sequence, χ will take a positive value. Otherwise, it will be negative as for the sub-triangle Δim_kc_{k-1}. Thus, Eq. (2.49) can be written as

$$A_{i,2k} = \frac{\chi}{2} \qquad (2.50)$$

Analogously, for the $(2k-1)$th sub-triangle, Δim_kc_{k-1}, the relevant coefficients are

$$b_{i,2k-1} = -a_{i,2k-1}\frac{(y_{c_{k-1}} - y_{m_k})}{\chi} \qquad (2.51)$$

$$c_{i,2k-1} = a_{i,2k-1}\frac{(x_{c_{k-1}} - x_{m_k})}{\chi} \qquad (2.52)$$

with

$$\chi = (x_{m_k} - x_i)(y_{c_{k-1}} - y_i) - (x_{c_{k-1}} - x_i)(y_{m_k} - y_i) \qquad (2.53)$$

being negative.

Thus, the relation between χ and the area of the sub-triangle Δim_kc_{k-1} becomes

$$A_{i,2k-1} = -\frac{\chi}{2} \tag{2.54}$$

2.2.3 *Approximation of spatial derivatives*

2.2.3.1 *First-order derivatives*

As mentioned in the preceding section, the pair of sub-triangles are considered together for the further formulations of LW-GSM. Correspondingly, Eq. (2.39) can be approximated as

$$\nabla U_i \approx -\sum_{k=1}^{n_i} \left[\int_{\Delta im_kc_k} U(\mathbf{x})\nabla\hat{w}(\mathbf{x}-\mathbf{x}_i)d\mathbf{x} \right.$$
$$\left. + \int_{\Delta im_kc_{k-1}} U(\mathbf{x})\nabla\hat{w}(\mathbf{x}-\mathbf{x}_i)d\mathbf{x} \right] \tag{2.55}$$

Substituting Eq. (2.43) into Eq. (2.55) yields

$$\frac{\partial U_i}{\partial x} \approx -\sum_{k=1}^{n_i} \left[b_{i,2k} \int_{\Delta im_kc_k} U(\mathbf{x})d\mathbf{x} + b_{i,2k-1} \int_{\Delta im_kc_{k-1}} U(\mathbf{x})d\mathbf{x} \right]$$
$$\tag{2.56}$$

$$\frac{\partial U_i}{\partial y} \approx -\sum_{k=1}^{n_i} \left[c_{i,2k} \int_{\Delta im_kc_k} U(\mathbf{x})d\mathbf{x} + c_{i,2k-1} \int_{\Delta im_kc_{k-1}} U(\mathbf{x})d\mathbf{x} \right]$$
$$\tag{2.57}$$

The integral for function $U(\mathbf{x})$ over each sub-triangle can be simply approximated in the fashion of

$$\int_{\Delta im_kc_k} U(\mathbf{x})d\mathbf{x} \approx \frac{1}{3}(U_i + U_{m_k} + U_{c_k})A_{i,2k} \tag{2.58}$$

$$\int_{\Delta im_kc_{k-1}} U(\mathbf{x})d\mathbf{x} \approx \frac{1}{3}(U_i + U_{m_k} + U_{c_{k-1}})A_{i,2k-1} \tag{2.59}$$

Substituting Eqs. (2.58) and (2.59) back into the Eqs. (2.56) and (2.57) gives

$$\frac{\partial U_i}{\partial x} \approx -\sum_{k=1}^{n_i} \left[\frac{1}{3}(U_i + U_{m_k} + U_{c_k})\xi_{i,2k} \right.$$

$$\left. + \frac{1}{3}(U_i + U_{m_k} + U_{c_{k-1}})\xi_{i,2k-1} \right] \qquad (2.60)$$

$$\frac{\partial U_i}{\partial y} \approx -\sum_{k=1}^{n_i} \left[\frac{1}{3}(U_i + U_{m_k} + U_{c_k})\eta_{i,2k} \right.$$

$$\left. + \frac{1}{3}(U_i + U_{m_k} + U_{c_{k-1}})\eta_{i,2k-1} \right] \qquad (2.61)$$

where

$$\begin{cases} \xi_{i,2k} = b_{i,2k}A_{i,2k} = -\frac{3}{2}\frac{y_{c_k} - y_{m_k}}{V_i} \\[2mm] \xi_{i,2k-1} = b_{i,2k-1}A_{i,2k-1} = \frac{3}{2}\frac{y_{c_{k-1}} - y_{m_k}}{V_i} \\[2mm] \eta_{i,2k} = c_{i,2k}A_{i,2k} = \frac{3}{2}\frac{x_{c_k} - x_{m_k}}{V_i} \\[2mm] \eta_{i,2k-1} = c_{i,2k-1}A_{i,2k-1} = -\frac{3}{2}\frac{x_{c_{k-1}} - x_{m_k}}{V_i} \end{cases} \qquad (2.62)$$

It should be mentioned those variables defined in Eq. (2.62) are derived by considering the relations between χ and the area of the respective sub-triangle. The adoption of $\xi_{i,2k}, \eta_{i,2k}, \xi_{i,2k-1}, \eta_{i,2k-1}$ in Eqs. (2.60) and (2.61) can avoid the involvement of the area for each sub-triangle and thus substantially save the memory for storing the area of each sub-triangle.

As implemented in Schemes VI and VIII, the relevant function values at midpoints of edges and centroids of triangles can be simply approximated with linear interpolation of values at constitutive nodes.

2.2.3.2 *Second-order derivatives*

Using linear interpolation in the integral calculation, the second-order derivatives can be approximated as follows:

$$
\frac{\partial^2 U_i}{\partial x^2} \approx -\sum_{k=1}^{n_i} \left[\frac{1}{3} \left(\frac{\partial U_i}{\partial x} + \frac{\partial U_{m_k}}{\partial x} + \frac{\partial U_{c_k}}{\partial x} \right) \xi_{i,2k} \right.
$$
$$
\left. + \frac{1}{3} \left(\frac{\partial U_i}{\partial x} + \frac{\partial U_{m_k}}{\partial x} + \frac{\partial U_{c_{k-1}}}{\partial x} \right) \xi_{i,2k-1} \right] \qquad (2.63)
$$

$$
\frac{\partial^2 U_i}{\partial y^2} \approx -\sum_{k=1}^{n_i} \left[\frac{1}{3} \left(\frac{\partial U_i}{\partial y} + \frac{\partial U_{m_k}}{\partial y} + \frac{\partial U_{c_k}}{\partial y} \right) \eta_{i,2k} \right.
$$
$$
\left. + \frac{1}{3} \left(\frac{\partial U_i}{\partial y} + \frac{\partial U_{m_k}}{\partial y} + \frac{\partial U_{c_{k-1}}}{\partial y} \right) \eta_{i,2k-1} \right] \qquad (2.64)
$$

In the above equations, the gradients at midpoints of edges and centroids of triangles can be approximated using either linear interpolation or gradient smoothing operations, as done in GSM with a constant smoothing function. As for the approximation with gradient smoothing operation, both constant and linear smoothing functions can be used. Thus, the following three ways are proposed to approximate the gradients at those non-storage locations and have been tested in this study:

1. simple linear interpolation of gradients at respective constitutive nodes;
2. gradient smoothing operators using constant smoothing functions;
3. gradient smoothing operators using piecewise linear smoothing functions.

The numerical results will be presented in Chapter 3 in detail.

2.3 Relationships between GSM and LW-GSM

The LW-GSM and the standard GSM differ in the type of smoothing functions they use, which are piecewise linear and constant, respectively. Due to this difference, the LW-GSM approximates partial differential derivatives by integrating over the surface of an nGSD, while the standard GSM integrates over the boundary of an nGSD. Table 2.4 summarizes the variations between the two methods.

Table 2.4. Comparison of GSM and LW-GSM.

	GSM	LW-GSM
Smoothing domains	nGSD, mGSD, cGSD	nGSD, mGSD, cGSD
Smoothing functions	$W = \begin{cases} 1/V_i, & \mathbf{x} \in \Omega_i^s \\ 0, & \mathbf{x} \notin \Omega_i^s \end{cases}$	$W = \begin{cases} \mathbf{a}_0 + \mathbf{a}_1\,(\mathbf{x} - \mathbf{x}_i), & \mathbf{x} \in \Omega_i^s \\ 0, & \mathbf{x} \notin \Omega_i^s \end{cases}$
Approximation of derivatives	$\begin{cases} \nabla U_i \approx \frac{1}{V_i} \oint_{\partial\Omega_i^s} U \mathbf{n} dS \\ \nabla \cdot (\nabla U_i) \approx \frac{1}{V_i} \oint_{\partial\Omega_i^s} \mathbf{n} \cdot \nabla U dS \end{cases}$	$\begin{cases} \nabla U_i \approx - \oint_{\Omega_i^s} U \nabla W dV \\ \nabla \cdot (\nabla U_i) \approx - \oint_{\Omega_i^s} U \cdot \nabla W dV \end{cases}$

Apparently, the two methods seem remarkably different from each other. However, once the simple linear approximation to the surface integral is adopted in the LW-GSM, the relation between the two methods can be identified.

2.3.1 *For interior nodes*

Substitution of Eq. (2.62) into Eqs. (2.60) and (2.61) generates the following:

$$
\begin{aligned}
\frac{\partial U_i}{\partial x} &\approx -\sum_{k=1}^{n_i}\left[-\frac{1}{3}\frac{3}{V_i}\frac{y_{c_k}-y_{m_k}}{2}(U_i+U_{m_k}+U_{c_k}) \right. \\
&\left. + \frac{1}{3}\frac{3}{V_i}\frac{y_{c_{k-1}}-y_{m_k}}{2}(U_i+U_{m_k}+U_{c_{k-1}}) \right] \\
&= \frac{1}{V_i}\sum_{k=1}^{n_i}\left[(y_{c_k}-y_{m_k})\left(\frac{U_i}{2}+\frac{U_{m_k}+U_{c_k}}{2} \right) \right. \\
&\left. + (y_{m_k}-y_{c_{k-1}})\left(\frac{U_i}{2}+\frac{U_{m_k}+U_{c_{k-1}}}{2} \right) \right] \\
&= \frac{1}{V_i}\sum_{k=1}^{n_i}\left[(y_{c_k}-y_{m_k})\left(\frac{U_{m_k}+U_{c_k}}{2} \right) \right. \\
&\left. + (y_{m_k}-y_{c_{k-1}})\left(\frac{U_{m_k}+U_{c_{k-1}}}{2} \right) \right]
\end{aligned}
\tag{2.65}
$$

$$\frac{\partial U_i}{\partial y} \approx -\sum_{k=1}^{n_i}\left[-\frac{1}{3}\frac{3}{V_i}\frac{x_{c_k}-x_{m_k}}{2}(U_i+U_{m_k}+U_{c_k})\right.$$

$$\left.+\frac{1}{3}\frac{3}{V_i}\frac{x_{c_{k-1}}-x_{m_k}}{2}(U_i+U_{m_k}+U_{c_{k-1}})\right]$$

$$=\frac{1}{V_i}\sum_{k=1}^{n_i}\left[(x_{c_k}-x_{m_k})\left(\frac{U_i}{2}+\frac{U_{m_k}+U_{c_k}}{2}\right)\right.$$

$$\left.+(x_{m_k}-x_{c_{k-1}})\left(\frac{U_i}{2}+\frac{U_{m_k}+U_{c_{k-1}}}{2}\right)\right]$$

$$=\frac{1}{V_i}\sum_{k=1}^{n_i}\left[(x_{c_k}-x_{m_k})\left(\frac{U_{m_k}+U_{c_k}}{2}\right)\right.$$

$$\left.+(x_{m_k}-x_{c_{k-1}})\left(\frac{U_{m_k}+U_{c_{k-1}}}{2}\right)\right] \tag{2.66}$$

Thus, the LW-GSM results in the same formulations as those obtained with the GSM Scheme VIII.

2.3.2 *For boundary nodes*

For Dirichlet boundary conditions, the values of the function on the boundary are known, and therefore, no further calculations are needed to update their values using the gradient smoothing operator.

However, when other types of boundary conditions are imposed, the values of the function at the boundary nodes need to be numerically solved to satisfy the governing equations similar to the inner nodes. In such cases, the gradients of the scalar quantity at the boundary nodes are approximated based on the given boundary conditions.

For instance, in the half-model case in the third Poisson problem in Chapter 3, the first-order derivative of the unknown quantity with respect to the x-axis is zero, which is directly used in the approximation of discretized governing equations at relevant boundary nodes. In this case, the gradients of the scalar U at the symmetric boundary, as illustrated in Figure 2.9, are approximated as

$$\frac{\partial U_i}{\partial x} = 0 \tag{2.67}$$

$$\frac{\partial U_i}{\partial y} \approx -\sum_{k=1}^{3}\left[\frac{1}{3}(U_i + U_{m_k} + U_{c_k})\eta_{i,2k}\right]$$

$$-\sum_{k=2}^{4}\left[\frac{1}{3}(U_i + U_{m_k} + U_{c_{k-1}})\eta_{i,2k-1}\right]$$

$$= -\sum_{k=1}^{3}\left[\frac{1}{3}\frac{3}{V_i}\frac{x_{c_k} - x_{m_k}}{2}(U_i + U_{m_k} + U_{c_k})\right]$$

$$-\sum_{k=2}^{4}\left[\frac{1}{3}\frac{3}{V_i}\frac{x_{c_{k-1}} - x_{m_k}}{2}(U_i + U_{m_k} + U_{c_{k-1}})\right]$$

$$= \frac{U_i}{2V_i}\left[(x_{m_1} - x_{c_1}) + (x_{m_2} - x_{c_2}) + (x_{m_3} - x_{c_3})\right.$$

$$\left. + (x_{c_1} - x_{m_2}) + (x_{c_2} - x_{m_3}) + (x_{c_3} - x_{m_4})\right]$$

$$+ \frac{1}{V_i}\sum_{k=1}^{3}\left[(x_{m_k} - x_{c_k})\frac{U_{m_k} + U_{c_k}}{2}\right]$$

$$+ \frac{1}{V_i}\sum_{k=2}^{4}\left[(x_{c_{k-1}} - x_{m_k})\frac{U_{m_k} + U_{c_{k-1}}}{2}\right]$$

$$= \frac{U_i}{2V_i}\left[(x_{m_1} - x_{m_4})\right] + \frac{1}{V_i}\sum_{k=1}^{3}\left[(x_{m_k} - x_{c_k})\frac{U_{m_k} + U_{c_k}}{2}\right]$$

$$+ \frac{1}{V_i}\sum_{k=2}^{4}\left[(x_{c_{k-1}} - x_{m_k})\frac{U_{m_k} + U_{c_{k-1}}}{2}\right]$$

$$= \frac{1}{V_i}\sum_{k=1}^{3}\left[(x_{m_k} - x_{c_k})\frac{U_{m_k} + U_{c_k}}{2}\right]$$

$$+ \frac{1}{V_i}\sum_{k=2}^{4}\left[(x_{c_{k-1}} - x_{m_k})\frac{U_{m_k} + U_{c_{k-1}}}{2}\right] \tag{2.68}$$

Again, the obtained formulations are identical to those in the GSM scheme VIII.

Similarly, we can also derive that the approximated second-order derivative (Laplace operator) has the same formulations as that in the GSM scheme VIII.

Hence, upon the above-mentioned derivation, we can conclude that the proposed LW-GSM scheme is equivalent to the GSM Scheme VIII. Such a finding has been verified in LW-GSM solutions to the Poisson equation and will be illustrated in detail in the following chapter.

2.4   A note on higher-order GSMs

The GSM can be extended to higher-order schemes, but this has not yet been done so far. This is still an open field of study for developing possible higher-order methods, and the GSM offers a good window of opportunity. One possible way is to use higher-order smoothing functions similar to the ones used in SPH [6]. Such smoothing functions are all designed to make the function and even its derivatives vanish on the smoothing domain boundary, and hence only domain integration is required similar to the LW-GSM. When a higher-order smoothing function is used, one may also increase the order of approximation for the field variables. The alternative is to construct higher-order smoothed derivatives of functions based on the so-called pick out theory developed recently [15]. This theory allows us to pick out any higher order approximation for the gradient. This high-order theory has been successfully applied to W2 formulations for smoothed finite element methods [16, 17].

References

[1]   G. R. Liu, *Meshfree Methods: Moving beyond the Finite Element Method*. Boca Raton, FL: CRC Press (2010).

[2]   S. C. Wu, G. R. Liu, H. O. Zhang, X. Xu, and Z. R. Li, *A Node-Based Smoothed Point Interpolation Method (NS-PIM) for Three-Dimensional Heat Transfer Problems*. Elsevier Masson (2009). doi: 10.1016/j.ijthermalsci.2008.10.010.

[3]   G. R. Liu, T. Nguyen-Thoi, H. Nguyen-Xuan, and K. Y. Lam, A node-based smoothed finite element method (NS-FEM) for upper bound solutions to solid mechanics problems. *Computers & Structures*, 87(1–2), 14–26 (2009, January). doi: 10.1016/j.compstruc.2008.09.003.

[4]   G. R. Liu *et al.*, A gradient smoothing method (GSM) with directional correction for solid mechanics problems. *Computational Mechanics*, 41(3), 457–472 (2008).

[5]   J. S. Chen, C. T. Wu, and T. Belytschko, Regularization of material instabilities by meshfree approximations with intrinsic length scales. *International Journal for Numerical Methods in Engineering*, 47, 1303–1322 (2000).

[6]   G. R. Liu, and M. B. Liu, *Smoothed Particle Hydrodynamics: A Meshfree Particle Method*. Singapore: World Scientific (2003).

[7] X. Xu, Development of gradient smoothing method (GSM) for fluid flow problems. National University of Singapore (2009). doi: https://scholarbank.nus.edu.sg/handle/10635/160959.

[8] G. R. Liu, and G. X. Xu, A gradient smoothing method (GSM) for fluid dynamics problems. *The International Journal for Numerical Methods in Fluids*, 58(10), 1101–1133 (2008, December).

[9] P. I. Crumpton, P. Moinier, and M. B. Giles, An unstructured algorithm for high Reynolds number flows on highly-stretched grids. In: *Proceedings of the 10th International Conference on Numerical Methods in Laminar and Turbulent Flows*, Swansea, England (1997).

[10] S. V. Patankar, *Numerical Heat Transfer and Fluid Flow.* New York: CRC Press; McGraw-Hill (1980). doi: 10.1201/9781482234213.

[11] T. J. Barth, Numerical methods for conservation laws on structured and unstructured meshes. Technical Report, VKI Lecture Series 2003 (2003).

[12] P. Moinier, Algorithm developments for an unstructured viscous flow solver. PhD Thesis, University of Oxford (1999).

[13] G. R. Liu, and G. X. Xu, A gradient smoothing method (GSM) for fluid dynamics problems. *International Journal for Numerical Methods in Fluids*, 58(10), 1101–1133 (2008, December). doi: 10.1002/fld.1788.

[14] E. Li, V. Tan, G. X. Xu, G. R. Liu, and Z. C. He, A novel linearly-weighted gradient smoothing method (LWGSM) in the simulation of fluid dynamics problem. *Computers & Fluids*, 50(1), 104–119 (2011, November).

[15] G. R. Liu, A novel pick-out theory and technique for constructing the smoothed derivatives of functions for numerical methods, *International Journal of Computational Methods*, 15(03), 1850070 (2018).

[16] X. Cui, S. Y. Duan, S. H. Huo, and G. R. Liu, A high order cell-based smoothed finite element method using triangular and quadrilateral elements. *Engineering Analysis with Boundary Elements*, 128, 133–148 (2021).

[17] Y. Li, G. R. Liu, Z. Feng, K. Ng, and S. Li, A node-based smoothed radial point interpolation method with linear strain fields for vibration analysis of solids. *Engineering Analysis with Boundary Elements*, 114, 8–22 (2020).

Chapter 3

Eulerian GSM for Solving Poisson Equation

Contents

Chapter 2 showed that GSM will work well for unstructured grids. This chapter presents numerical examinations on the efficacy and performance of various GSM schemes. It is applied to solve a number of benchmarking problems: Poisson equations with different known source terms to which we have analytical solutions for objective assessments. Our discussion progresses as follows:

- Section 3.1 presents the settings of three benchmarking problems.
- Section 3.2 defines the error indexes for assessing numerical solutions.

- Section 3.3 presents various types of cells used to obtain numerical solutions.
- Section 3.4 discusses the standard GSM solutions that use the constant smoothing function.
- Section 3.5 discusses the solution obtained using the linear gradient smoothing function, known as the Linear Weighted Gradient Smoothing Method (LW-GSM).
- Section 3.6 summarizes the findings.

Based on the stencil analyses in Chapter 2, four schemes (II, VI, VII, and VIII) are selected and implemented in our GSM solver, because of their stencil compactness with favorable coefficients. Schemes II and VI are partial GSM because they use linear interpolation and direction correction. Schemes VII and VIII are full GSM, where all the approximation of gradient is done only by gradient smoothing and no direction correction is needed.

The numerical results will be for two-dimensional Poisson equations solved using GSM with these four schemes. Comparison studies are conducted in terms of numerical accuracy and computational efficiency. The roles of directional correction and gradient smoothing technique used for the approximation of the gradient at midpoints of cell edges are numerically studied. In addition, the effects of the shape, the density, and the irregularity of cells upon the accuracy and stability are also investigated.

3.1　Problem description

The Poisson equation governs many real-life problems in sciences and engineering, such as diffusion problems of groundwater flow, heat transfer with heat sources, gravitation, and electrostatics. It is an elliptic PDE and a generalization of Laplace's equation.

This section solves the Poisson equation defined in a 2D square domain:

$$\frac{\partial U}{\partial t} = \frac{\partial^2 U}{\partial x^2} + \frac{\partial^2 U}{\partial y^2} - f(x,y,t), \quad (0 < x < 1, 0 < y < 1) \tag{3.1}$$

where f is a given source term.

Dirichlet conditions are applied to the boundaries, i.e., the values of field functions at the boundaries are prescribed. The pseudo-transient approach is adopted for pursuing steady-state solutions, using the GSM with a dual time-stepping technique (to be discussed in detail in Section 4.2.3). Three problems with variations of source and boundary conditions are studied.

For the **first** Poisson problem, the source and initial conditions are given as

$$\left. \begin{array}{l} f(x,y,t) = 13\,exp(-2x+3y) \\ U(x,y,0) = 0 \end{array} \right\}, \quad (0 < x < 1, 0 < y < 1) \qquad (3.2)$$

As plotted in Figure 3.1(a), the analytical solution to this problem is known and given as

$$\widehat{U}(x,y) = e^{(-2x+3y)}, \quad (0 < x < 1, 0 < y < 1) \qquad (3.3)$$

For the **second** Poisson problem, the source, initial conditions, and analytical solution are given as

$$\left. \begin{array}{l} f(x,y,t) = \sin(\pi x)\sin(\pi y) \\ U(x,y,0) = 0 \end{array} \right\}, \quad (0 < x < 1, 0 < y < 1) \qquad (3.4)$$

$$\widehat{U}(x,y) = -\frac{1}{2\pi^2}\sin(\pi x)\sin(\pi y), \quad (0 < x < 1, 0 < y < 1) \qquad (3.5)$$

The contour plot of the analytical solution to the second problem is shown in Figure 3.1(b).

The **third** Poisson problem is identical to the second one, but only covers half of the domain $(0 < x < 0.5, 0 < y < 1)$, as shown in Figure 3.1(c), considering the symmetry of the second problem with respect to the y-axis. A symmetry boundary condition is imposed on the right side vertical bounding face, i.e.,

$$\frac{\partial U}{\partial x}(x = 0.5) = 0 \qquad (3.6)$$

The Dirichlet boundary conditions are imposed onto the remaining three bounding faces of the halved domain. This problem is purposely designed to evaluate the GSM and LW-GSM solvers for problems with mixed types of boundary conditions.

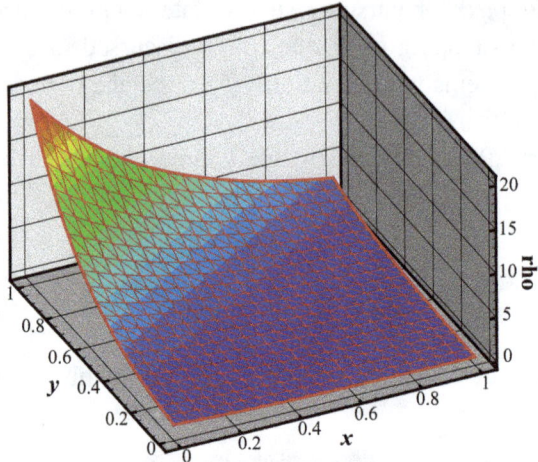

(a) The **first** Poison problem

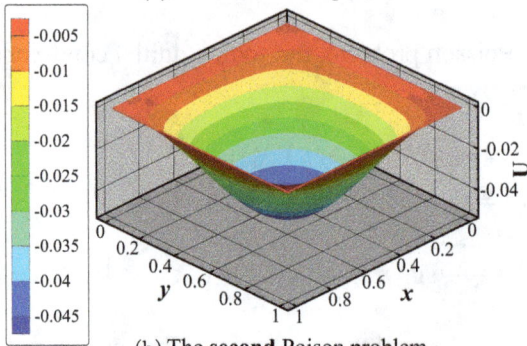

(b) The **second** Poison problem

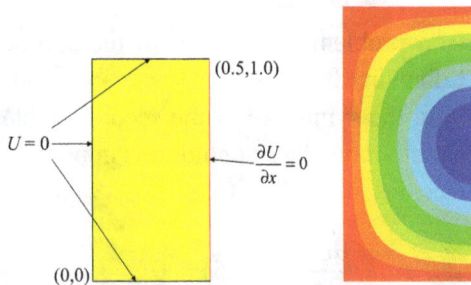

(c) The **third** Poison problem

Figure 3.1. Contour plots of exact solutions to three Poisson problems (Reprinted with permission from [1]).

3.2 Evaluation of numerical errors

Three error indexes are used in this study. The **first** is the convergence error index, ε_{con}, defined as

$$\varepsilon_{con} = \frac{\sqrt{\sum_{i=1}^{n_{node}} (U_i^{(n+1)} - U_i^{(n)})^2}}{\sqrt{\sum_{i=1}^{n_{node}} (U_i^{(1)} - U_i^{(0)})^2}} \tag{3.7}$$

where $U_i^{(n)}$ denotes the numerical value of the field variable at node i at the n-th iteration, and n_{node} is the total number of nodes in the domain. The value of ε_{con} is monitored during iterations and used to terminate the iterative process. In most simulations, in order to exclude the effect due to the temporal discretization, computations continue until ε_{con} becomes small and stabilized, as indicated in Figure 3.2.

The **second** is the L_2-norm of the error, defined by

$$error = \frac{\sqrt{\sum_{i=1}^{n_{node}} (U_i - \widehat{U}_i)^2}}{\sqrt{\sum_{i=1}^{n_{node}} (\widehat{U}_i)^2}} \tag{3.8}$$

where U_i and $\widehat{U}_i$ are numerical and analytical solutions at node i, respectively. This error index is used to examine the accuracy among different schemes.

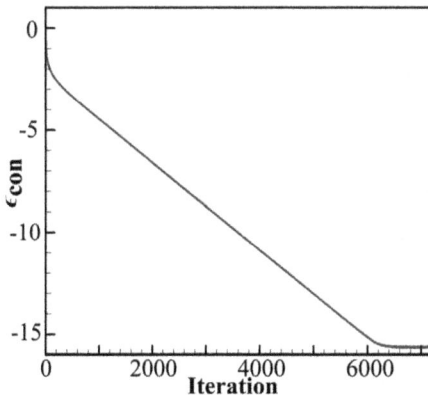

Figure 3.2. Typical convergence curve profile (Reprinted with permission from [1]).

The **third** is the node-wise relative error:

$$r\_error_i = \frac{|U_i - \widehat{U}_i|}{|\widehat{U}_i|} \tag{3.9}$$

This gives the distribution of the error over the domain and is used to identify problematic regions of the numerical solution.

3.3 Cell types used in study

This section evaluates the adaptability of GSM to different cell types, including square, right triangle, regular triangle, and irregular triangle cells, as shown in Figure 3.3. The numerical results for each cell type and different spatial discretization schemes are discussed in detail, providing

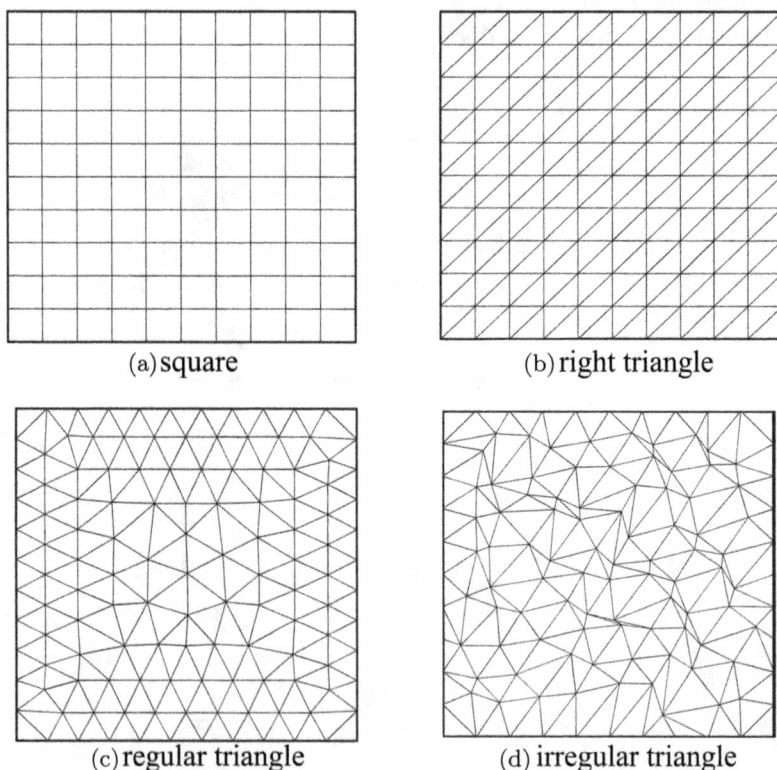

(a) square   (b) right triangle

(c) regular triangle   (d) irregular triangle

Figure 3.3.   Cell types used in the study (Reprinted with permission from [1]).

valuable insights to the performance of GSM under various discretization conditions.

3.4 GSM solutions

3.4.1 *The role of directional correction*

Scheme I (see Chapter 2) is found to be inadequate for solving the first Poisson problem using square-shaped cells, because it generates saw-toothed numerical errors (checkerboard problem), as shown in Figure 3.4(a). Scheme II, which is Scheme I with directional correction, effectively overcomes this issue, as shown in Figure 3.4(b). This numerical finding is consistent with the stencil analysis performed in Chapter 2.

The use of directional correction also significantly reduces the overall numerical error, as shown in Table 3.1, where the error magnitudes decrease by about five times compared to Scheme I. However, Scheme II requires a smaller time step for stability requirements of the CFL condition, resulting in more computations compared to Scheme I, especially when using fine meshes.

3.4.2 *Comparison of four favorable schemes*

The L_2 errors for the first Poisson problem is plotted in Figures 3.5–3.7 against the averaged node spacing for, respectively, the square cells, right

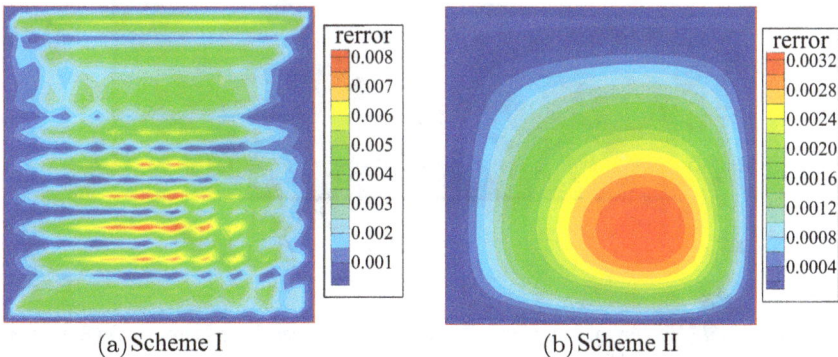

(a) Scheme I   (b) Scheme II

Figure 3.4. Contour plots of node-wise relative errors using square cells for the first Poisson problem (Reprinted with permission from [1]).

Table 3.1. Comparison of numerical errors with Scheme I and Scheme II for the first Poisson problem [1].

	Scheme I		Scheme II	
No. of nodes	Error	Iteration	Error	Iteration
36	1.96E-002	20	5.07E-003	24
121	8.58E-003	89	1.66E-003	82
441	2.63E-003	202	4.87E-004	303
1681	7.16E-004	728	1.33E-004	1379
6561	1.86E-004	2679	3.38E-005	4299

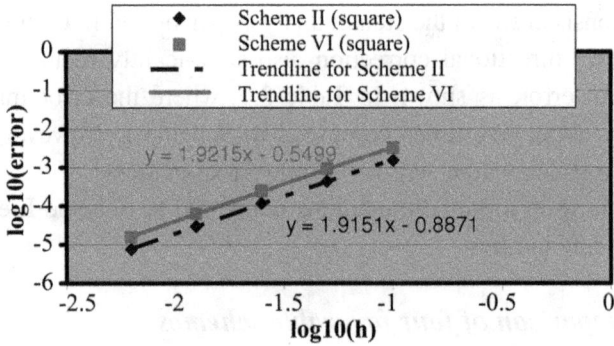

(a)

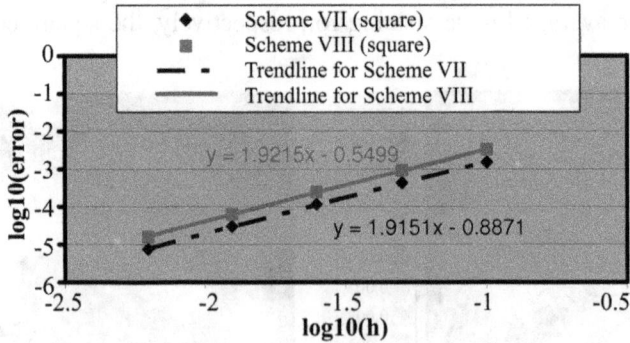

(b)

Figure 3.5. L_2 numerical error against averaged node spacing (square cells) (Reprinted with permission from [1]).

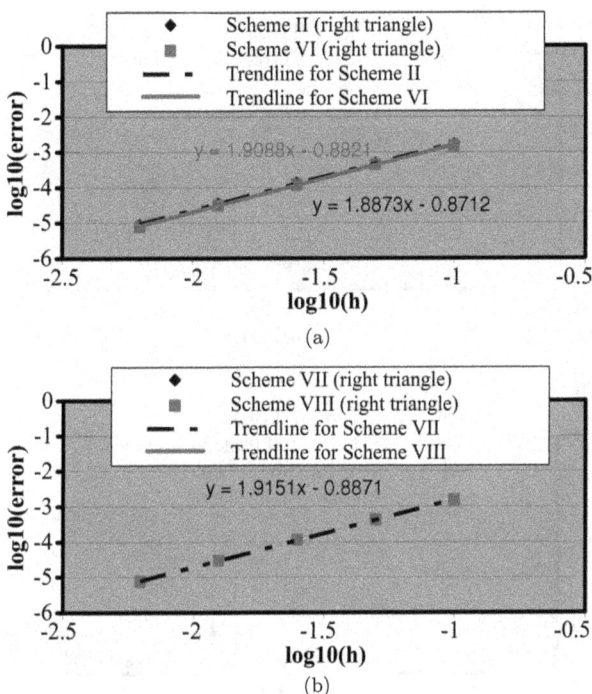

Figure 3.6. L_2 numerical error against averaged node spacing (right triangular cells) (Reprinted with permission from [1]).

triangular cells, and regular triangular cells. The errors are evaluated using Eq. (3.8), and the averaged node spacing, h, is calculated using [2]

$$h = \frac{V}{\sqrt{n_{node}} - 1} \tag{3.10}$$

where V denotes the area of the whole computational domain. This formula states that the averaged node spacing inversely depends largely on the square root of the number of nodes for 2D domains.

On square cells

From Figures 3.5(a) and 3.5(b), it is observed that scheme VII as accurate as scheme II. These two schemes have the same discretization stencils as previously analyzed. The same holds true for schemes VI and VIII. However, the two-point quadrature schemes (VI and VIII) have lower accuracy and higher computational costs compared to the one-point quadrature

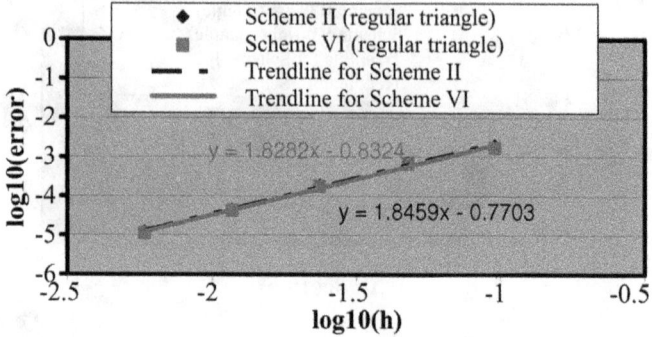

(a)

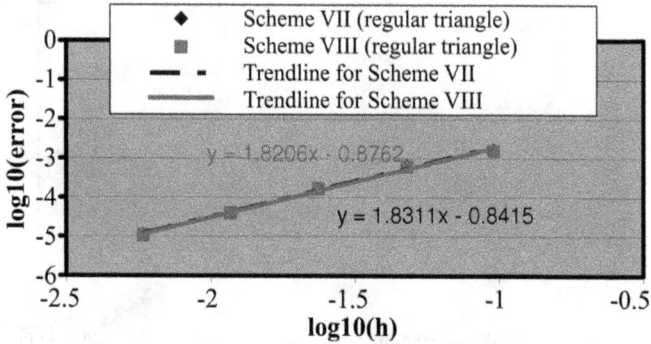

(b)

Figure 3.7. L_2 numerical error against averaged node spacing (regular triangular cells) (Reprinted with permission from [1]).

schemes (II and VII). Thus, schemes II and VII are found to be slightly better performers than schemes VI and VIII for square-shaped cells. The slope coefficients of the trend lines further verify the 2nd-order accuracy of all four schemes, as previously determined from the analysis of truncation errors.

On right triangular cells
Numerical error against averaged node spacing for right triangular cells is shown in Figures 3.6(a) and 3.6(b). It is seen that Scheme VI gives slightly more accurate results than Scheme II. Scheme VII is as accurate as Scheme VIII.

On cells in regular triangle shape

For regular triangular cells (Figure 3.3(c)), both two-point quadrature schemes (VI and VIII) are slightly more accurate than one-point quadrature schemes, as shown in Figures 3.7(a) and 3.7(b). This is consistent with the findings for the right triangular cells.

Table 3.2 summarizes the numerical errors for different discretization schemes when regular triangular cells are used. Scheme VII is slightly more accurate than Scheme II, while Scheme VIII is more accurate than Scheme VI. This is also true when the right triangular cells are used in the simulations. The discrepancy in accuracy is related to the approximation of the gradient at boundary nodes. Schemes II and VI introduce additional errors because they require approximating the gradient at boundary nodes when estimating the gradient at midpoints of internal cell edges connected to those boundary nodes. Schemes VII and VIII, subjected to Dirichlet boundary conditions, avoid such approximations at boundary nodes using the gradient smoothing techniques *via* mGSDs, resulting in a full GSM method.

In terms of computational cost, Schemes VII and II are comparable. The same is true for Schemes VIII and VI. Schemes VI and VIII require roughly twice the computational time of Schemes II and VII. While Schemes VI and VIII are more accurate than Schemes II and VII, the improvement in accuracy is not significant for these cell types used. Therefore, for a balance between numerical accuracy and computational efficiency, the two one-point quadrature schemes (II and VII) are preferred in practice.

Table 3.2. Comparison of L_2 numerical errors using the regular triangle cells for the first Poisson problem [1].

		Numerical error			
No. of nodes	Time-step	Scheme II	Scheme VI	Scheme VIII	Scheme VIII
131	8.00E-3	1.95E-003	1.68E-3	1.65E-003	1.55E-003
478	1.00E-3	7.02E-004	6.46E-4	6.53E-004	6.20E-004
1887	5.00E-4	1.89E-004	1.74E-4	1.70E-004	1.65E-004
7457	1.00E-4	4.38E-005	4.12E-5	3.98E-005	3.94E-005
29629	3.00E-5	1.22E-005	1.11E-5	1.10E-005	1.05E-005

3.4.3 *Robustness to irregularity of cells*

The consistent findings described in the previous sections are also observed in the solutions to the second Poisson problem. Additional studies on effects of irregularity of triangular cells are carried out. It is well known that triangular cells have the best adaptability to complex geometries and can be generated automatically in very efficient ways. Therefore, it is desirable to use triangular cells for engineering problems with complicated geometries. Our objective is thus to further identify the sensitivity of the GSM to cell irregularity. The two one-point quadrature schemes (II and VII) are used in this study.

To study this in a systematic manner, we first define the irregularity for all triangular cells in the computational domain, γ, using the following formula:

$$\gamma = \frac{\sum_{i=1}^{n_e} \frac{(a_i-b_i)^2+(b_i-c_i)^2+(c_i-a_i)^2}{a_i^2+b_i^2+c_i^2}}{n_e} \tag{3.11}$$

where $a_i, b_i,$ and c_i, denote the lengths of the three cell edges of a triangular cell and n_e stands for the total number of cells in the overall domain. Equation (3.11) is derived from the formula proposed by Stillinger *et al.* [3] for a single triangle. Using Eq. (3.11), the irregularity vanishes for equilateral triangles and is positive for all other shapes including isosceles triangles.

Figure 3.8 shows six sets of triangular cells with various irregularities, but the same number of nodes. It is obvious that as the irregularity increases, the mesh is distorted further.

When the irregularity is larger than 0.152, overlapped cells are found in the domain, as shown in Figure 3.9, which is an extreme case that will not normally happen in practice. Such an extreme case cannot be accurately reflected in the irregularity using Eq. (3.11). Nevertheless, we still use such extremely distorted cells to test the robustness of the GSM.

Convergent results for all irregular meshes are obtained using the GSM with Schemes II and VII. Contour plots of the predicted solutions with Scheme VII are selectively shown in Figure 3.10. It is clear that the predicted results are reasonably accurate on all sets of irregular cells. Note that as irregularity of cells increases, the time step $(\Delta\tau)$ must be reduced so as to guarantee stable and convergent results, as shown in Table 3.3.

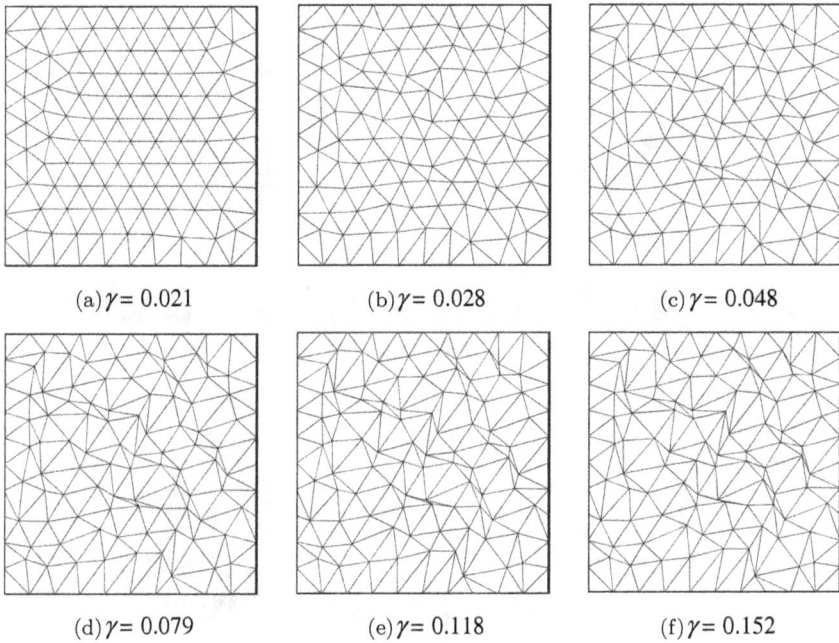

(a) $\gamma = 0.021$ (b) $\gamma = 0.028$ (c) $\gamma = 0.048$

(d) $\gamma = 0.079$ (e) $\gamma = 0.118$ (f) $\gamma = 0.152$

Figure 3.8. Triangular cells with various irregularities (Reprinted with permission from [1]).

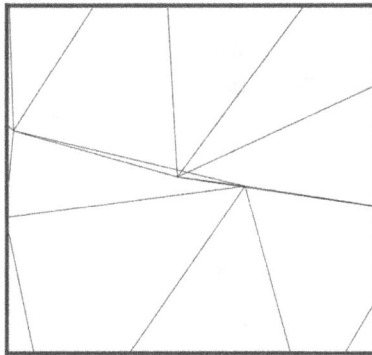

Figure 3.9. Overlapped cells in the computational domain ($\gamma = 0.16$), a case of extremely distorted cells (Reprinted with permission from [1]).

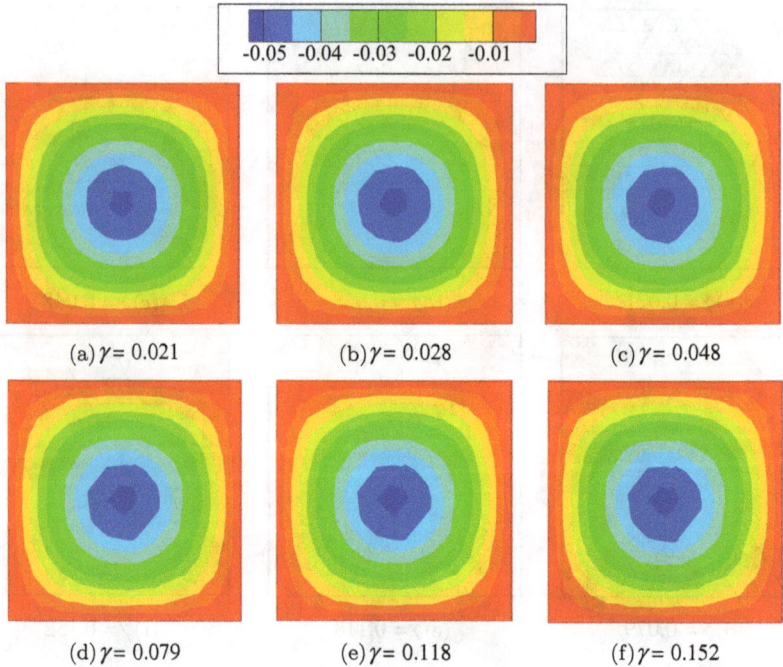

Figure 3.10. Contour plots of numerical solutions to the second Poisson equation (irregular triangular cells) (Reprinted with permission from [1]).

Table 3.3. Comparison of allowable maximum time step and L_2 numerical error when irregular triangular cells are used [1].

Mesh	Irregularity	Maximum time-step ($\Delta\tau$)	Error Scheme VII	Scheme II
(a)	0.0210	0.0100	0.0163	0.0172
(b)	0.0280	0.0100	0.0169	0.0177
(c)	0.0480	0.0090	0.0179	0.0188
(d)	0.0790	0.0075	0.0194	0.0206
(e)	0.1180	0.0040	0.0214	0.0231
(f)	0.1520	0.0005	0.0234	0.0259

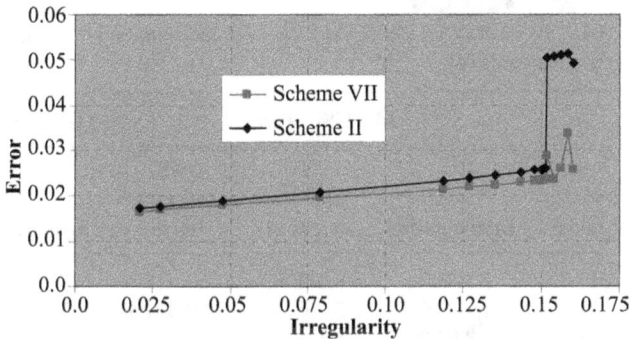

Figure 3.11. L_2 numerical errors in GSM solutions (Scheme II and VII) to the second Poisson problem with respect to the irregularity of cells (Reprinted with permission from [1]).

One finds that for cases without overlapped cells, the numerical errors predicted with the GSM do not vary so much, as the irregularity of cells increases (see Figure 3.11). Once overlapped cells occur in the domain, sudden jumps in numerical errors are observed. However, for both Schemes II and VII, stable results are still obtained. Besides, Scheme VII shows much better stability and accuracy than Scheme II among all irregular cells examined here. This means that the full GSM with Scheme VII is remarkably robust and insensitive to cell irregularity. In other words, with the proposed GSM, stable and accurate results can be obtained even with highly distorted triangular cells. Such an attractive feature can be attributed to the consistent use of smoothing operations in Scheme VII, which provides the crucial stability and robustness to our GSM.

3.5 LW-GSM solutions

The linearly weighted gradient smoothing method (LW-GSM), as described in Chapter 2, is tested for the second and third Poisson problems. The triangular cells, including right and irregular triangles, are primarily used in these tests. As mentioned in Chapter 2, there are three ways to approximate the gradient at non-storage locations: simple linear interpolation, GS operator using piecewise constant smoothing function, and GS operator using piecewise linear smoothing function. Comparison studies have been conducted, and the findings are summarized in the following.

Table 3.4. L_2 errors with LW-GSM for the second Poisson problem, using right triangular cells [4].

	LW-GSM error			GSM error
No. of nodes	Linear Interpolation	GS (piecewise constant)	GS (piecewise linear)	Scheme VIII
11×11	7.30E-02	8.27E-03	8.27E-03	8.27E-03
21×21	1.78E-02	2.06E-03	2.06E-03	2.06E-03
41×41	4.44E-03	5.14E-04	5.14E-04	5.14E-04
81×81	1.11E-03	1.29E-04	1.29E-04	1.29E-04
161×161	2.78E-04	3.21E-05	3.21E-05	3.21E-05

Table 3.5. L_2 errors with LW-GSM for the second Poisson problem, using irregular triangular cells [4].

	LW-GSM error			GSM error
No. of nodes	Linear Interpolation	GS (piecewise constant)	GS (piecewise linear)	Scheme VIII
131	6.93E-02	9.96E-03	9.96E-03	9.96E-03
478	1.58E-02	3.31E-03	3.31E-03	3.31E-03
1887	3.32E-03	7.90E-04	7.90E-04	7.90E-04
7457	8.16E-04	1.95E-04	1.95E-04	1.95E-04
29629	1.97E-04	4.95E-05	4.95E-05	4.95E-05

3.5.1 *Second Poisson problem*

Tables 3.4 and 3.5 summarize the LW-GSM errors regarding the number of nodes adopted to discretize the problem domain, together with the errors of the GSM with Scheme VIII.

It is clear that the LW-GSM errors are effectively reduced as the number of nodes increases. The use of gradient smoothing operations for approximating gradients at non-storage locations always gives higher accuracy than the use of linear interpolation. Besides, the LW-GSM errors based on the consistent use of gradient smoothing operations at different locations are found to be identical for both smoothing functions. In addition, the corresponding LW-GSM errors are equal to those obtained in GSM Scheme VIII. This finding supports the theoretical proof given in Chapter 2 that LW-GSM and GSM are essentially identical.

3.5.2 Third Poisson problem

The LW-GSM has also been applied for solutions to the third Poisson problem. Only regular triangles are used in this study.

As summarized in Table 3.6, the consistent usage of gradient smoothing operators, regardless of the type of smoothing functions, gives identical LW-GSM errors. Those numerical errors are identical to the GSM errors with Scheme VIII, and all much smaller than those using simple linear interpolation for approximating gradients. The difference can also be found in Figure 3.12 for the distribution of relative errors across the field. The solutions using simple linear interpolation can lead to checkerboard problems, as shown in Figure 3.12(a). Such problems can be effectively avoided using smoothing operations at the non-storage locations, as illustrated in Figure 3.12(b). As a result, the total errors are substantially reduced, as indicated in Table 3.6 and Figure 3.13.

Figure 3.13 depicts the profile of the LW-GSM error versus the averaged nodal spacing. The LW-GSM achieves at least 2nd-order accuracy.

Table 3.6. L_2 errors with LW-GSM on irregular triangular cells used in the third Poisson problem [4].

	LW-GSM error			GSM error
No. of nodes	Linear Interpolation	GS (piecewise constant)	GS (piecewise linear)	Scheme VIII
21	2.80E-01	7.93E-02	7.93E-02	7.93E-02
72	6.00E-02	1.58E-02	1.58E-02	1.58E-02
244	1.46E-02	3.85E-03	3.85E-03	3.85E-03
963	4.06E-03	8.21E-04	8.21E-04	8.21E-04
3741	8.21E-04	1.99E-04	1.99E-04	1.99E-04

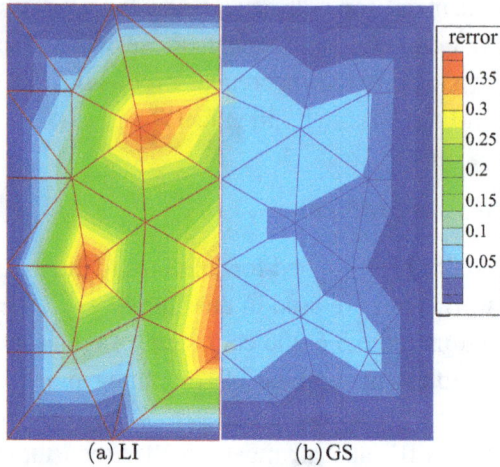

(a) LI   (b) GS

Figure 3.12. Contour plots of relative errors for the third Poisson problem (Reprinted with permission from [4]).

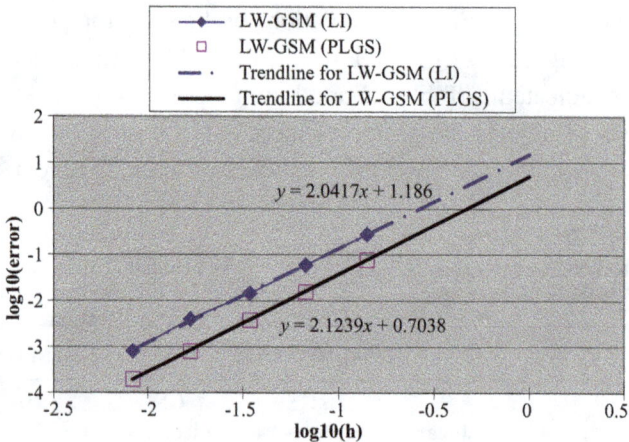

Figure 3.13. L_2 numerical errors in the LW-GSM solutions against the averaged nodal spacing (Reprinted with permission from [4]).

Besides, the algorithm based on consistent use of gradient smoothing operations at different locations gives a slightly higher convergence rate than that using the simple linear interpolation for an approximating gradient.

3.6 Remarks

In this study, schemes of a conservative gradient smoothing method formulated for the strong form of governing equations have been examined *via* numerical studies. The standard GSM is found to be effective for not only regular but also irregular cells, which provides flexibility to deal with fluid flow problems with complex geometry. We draw the following conclusions:

- Schemes II, IV, VI, and VIII are favorable, because of their compact stencils with positive coefficients of influence;
- Schemes VII and VIII are full GSM models that use the gradient smoothing technique on mGSDs and outperform Schemes II and VI in terms of robustness, stability, and accuracy;
- The one-point quadrature schemes (Schemes II and VII) have well-balanced performance in terms of both efficiency and accuracy;
- Scheme VII is superior to Scheme II in terms of consistency in derivative approximation at various locations and robustness against the irregularity of cells. Therefore, Scheme VII is most preferable in practice, especially for large-scale problems. It is a full GSM model without using any correction, using the minimum (one) integration point.
- Scheme VIII is also a full GSM model without using any correction, and is as good as Scheme VII in terms of accuracy, but it uses one extra point for the integration, and hence is a little expensive. Schemes VIII is expected to be more robust.

The LW-GSM is based on piecewise linear smoothing function. It is an alternate form of approximating the spatial derivatives. It is different in form, but identical to the GSM Scheme VIII, as verified in our numerical studies.

In conclusion, the standard GSM Scheme VII is the most preferable scheme and will be further explored for solutions to general complex fluid flow problems in the following chapters.

References

[1] G. R. Liu and G. X. Xu, A gradient smoothing method (GSM) for fluid dynamics problems, *International Journal for Numerical Methods in Fluids*, 58(10), 1101–1133 (2008, December).

[2] G. R. Liu, *Meshfree Methods: Moving Beyond the Finite Element Method*. Boca Raton, FL: CRC Press, (2010).

[3] D. K. Stillinger, F. H. Stillinger, S. Torquato, T. M. Truskett, and P. G. Debenedetti, Triangle distribution and equation of state for classical rigid disks, *Journal of Statistical Physics*, 100(1–2), 49–72 (2000, July). doi: 10.1023/A:1018675208867.

[4] X. Xu, Development of Gradient Smoothing Method (GSM) for fluid flow problems, National University of Singapore, (2009). doi: https://scholarbank.nus.edu.sg/handle/10635/160959.

Chapter 4

Eulerian GSM for Compressible Flows

Contents

In this chapter, we extend the application of GSM to time-dependent compressible flow problems governed by the Navier–Stokes equations in the

Eulerian frame. The chapter is organized as follows:

- Section 4.1 introduces the differential form of governing equations for compressible flows associated with the turbulence model employed in study.
- Section 4.2 presents the necessary numerical treatments of the proposed GSM framework when applied to solving time-dependent problems, including the spatial discretization, boundary treatment, and time integration.
- Section 4.3 presents the applications of GSM to a series of compressible flows, including the inviscid flow over the NACA0012 airfoil, laminar flow over the flat plate, turbulent flow over the RAE2822 airfoil, and time-dependent compressible laminar flow over a circular cylinder.

4.1　Setting of the problem

Fluid flows are governed by the well-known Navier–Stokes (N–S) equations, which describe the conservation of flow continuity, momentum, and energy in both compressible and incompressible flows. Our attention will be on the application of GSM for spatial discretization, both for compressible flows in this chapter and incompressible flows in the next chapter. We will present GSM algorithms for both transient and steady-state simulations, allowing for a comprehensive analysis of the GSM performance in solving complex fluid flow problems.

We start by introducing the differential form of the N–S equations, laying the foundation for subsequent discussions on the numerical treatments and applications of GSM in this chapter.

4.1.1　*Governing equations for compressible flows*

In the Cartesian coordinate system, without consideration of source terms, the governing N–S equation is a set of PDEs as follows:

Continuity equation:

$$\frac{\partial \rho}{\partial t} + \nabla \cdot (\rho \mathbf{v}) = 0 \tag{4.1}$$

where ρ is the density of fluid, t is the time, $\mathbf{v}$ is the velocity, and $\nabla \cdot$ represents the divergence operator.

Momentum equation:

$$\frac{\partial(\rho \mathbf{v})}{\partial t} + \nabla \cdot (\rho \mathbf{v} \otimes \mathbf{v}) = -\nabla p + \nabla \cdot \boldsymbol{\tau} \tag{4.2}$$

where p is the pressure, $\boldsymbol{\tau}$ is the stress tensor accounting for viscous effects, and $\mathbf{v} \otimes \mathbf{v}$ is the outer product with $\mathbf{u} \otimes \mathbf{v} = \mathbf{u}\mathbf{v}^{\mathrm{T}}$.

Energy equation:

$$\frac{\partial(\rho E_t)}{\partial t} + \nabla \cdot (\rho \mathbf{v} E_t) = -p\nabla \cdot \mathbf{v} + \nabla \cdot (\mathbf{q} + \mathbf{v} \cdot \boldsymbol{\tau}) \tag{4.3}$$

where E_t is the total energy per unit volume, and $\mathbf{q}$ is the heat flux as a function of temperature T and thermal conductivity k: $\mathbf{q} = -k\nabla T$.

The total internal energy E_t is defined as

$$E_t = e + \frac{1}{2}|\mathbf{v}|^2 = \frac{1}{\gamma}c_p T + \frac{1}{2}|\mathbf{v}|^2 \tag{4.4}$$

with

- e = internal energy
- c_p = specific heat at constant pressure
- γ = ratio of specific heat of the fluid media.

To solve the governing N-S equation, we need to specify the equation of state, stress tensor $\boldsymbol{\tau}$, dynamical viscosity μ, and thermal conductivity k.

Equation of state:
The equation of state for calorically perfect gas is used:

$$p = \rho RT \tag{4.5}$$

where R denotes the specific gas constant. The equation of state gives the expression to update the pressure with attained values for conservative variables. Considering the relation of $R = c_p - c_p/\gamma$ and Eq. (4.4), Eq. (4.5) for the update can be write as

$$p = (\gamma - 1)\rho \left[E_t - \frac{1}{2}|\mathbf{v}|^2 \right] \tag{4.6}$$

Stress tensor:

For Newtonian flows, it is reasonably assumed that the viscous stress is linearly proportional to the respective strain rate. The viscous stresses can then be written in terms of velocities:

$$
\begin{cases}
\tau_{xx} = 2\mu \dfrac{\partial u}{\partial x} + \lambda \nabla \cdot \mathbf{v} \\[2mm]
\tau_{yx} = \tau_{xy} = \mu \left(\dfrac{\partial u}{\partial y} + \dfrac{\partial v}{\partial x} \right) \\[2mm]
\tau_{yy} = \mu \dfrac{\partial v}{\partial y} + \lambda \nabla \cdot \mathbf{v}
\end{cases}
\tag{4.7}
$$

where

- λ = secondary viscous coefficient
- μ = dynamic viscosity coefficient
- $\mu \partial u / \partial x$ and $\mu \partial v / \partial y$ = rate of linear dilatation — a change in shape
- $\lambda \nabla \cdot \mathbf{v}$ = rate of volumetric dilatation, which reflects essentially a change in density.

For the normal stresses, the Stokes hypothesis can be used:

$$
\lambda + \frac{2}{3}\mu = 0
\tag{4.8}
$$

Except for extremely high temperature and pressure, this relation usually holds for typical fluid flow problems. Thus, the formulas for stresses can be simplified as

$$
\begin{cases}
\tau_{xx} = 2\mu \left(\dfrac{\partial u}{\partial x} - \dfrac{1}{3}\nabla \cdot \mathbf{v} \right) \\[2mm]
\tau_{yx} = \mu \left(\dfrac{\partial u}{\partial y} + \dfrac{\partial v}{\partial x} \right) \\[2mm]
\tau_{yy} = 2\mu \left(\dfrac{\partial v}{\partial y} - \dfrac{1}{3}\nabla \cdot \mathbf{v} \right)
\end{cases}
\tag{4.9}
$$

Dynamical viscosity:

For ideal gas flows, which will be focused on in this chapter, the dynamical viscosity is related to the temperature by the Sutherland law [1]:

$$\mu = \mu_\infty \left(\frac{T}{T_\infty} \right)^{3/2} \frac{T_\infty + 110}{T + 110} \tag{4.10}$$

in which the reference dynamic viscosity is calculated using

$$\mu_\infty = \frac{\rho_\infty |\mathbf{v}|_\infty L}{Re} \tag{4.11}$$

where

- ρ_∞ = reference density,
- $|\mathbf{v}|_\infty$ = reference speed,
- Re = the corresponding Reynolds number of the freestream flow,
- L = characteristic length of the problem domain of interest.

Thermal conductivity:

The thermal conductivity is estimated as

$$k = c_p \frac{\mu}{Pr} \tag{4.12}$$

where Pr denotes Prandtl number (0.72 for standard air).

4.1.2 Turbulence model

The one-equation Spalart–Allmaras model [2] is incorporated with the N–S equations to model turbulent flows. The turbulence governing equation with respect to kinematic turbulent viscosity $\tilde{v}$ takes the form of

$$\frac{\partial \tilde{v}}{\partial t} + \nabla \cdot (F_{c,T} - F_{v,T}) = S_T \tag{4.13}$$

where $F_{c,T}, F_{v,T}$, and S_T denote the turbulent convective flux, turbulent viscous flux, and relevant source terms, respectively.

For the turbulent field without concerns of transition, we have

$$\begin{cases} F_{c,T} = \tilde{v}\mathbf{v}\cdot\mathbf{n}, \quad F_{v,T} = n_x\tau_{xx}^T + n_y\tau_{yy}^T \\ S_T = C_{b1}\tilde{S}\tilde{v} + \dfrac{C_{b2}}{\sigma}\left[\left(\dfrac{\partial\tilde{v}}{\partial x}\right)^2 + \left(\dfrac{\partial\tilde{v}}{\partial y}\right)^2\right] - C_{w1}f_w\left(\dfrac{\tilde{v}}{d_w}\right)^2 \end{cases} \quad (4.14)$$

with

$$\tau_{yy}^T = \frac{1}{\sigma}(v_l + \tilde{v})\frac{\partial\tilde{v}}{\partial y} \quad (4.15)$$

$$\tau_{xx}^T = \frac{1}{\sigma}(v_l + \tilde{v})\frac{\partial\tilde{v}}{\partial x} \quad (4.16)$$

$$\tilde{S} = f_{v3}S + \frac{\tilde{v}}{\kappa^2 d_w^2}f_{v2} \quad (4.17)$$

$$f_w = g\left(\frac{1+C_{w3}^6}{g^6+C_{w3}^6}\right)^{1/6} \quad (4.18)$$

$\mathbf{n}$ is the unit normal vector with components of n_x, n_y in x and y directions, respectively. The subscript l denotes the laminar terms. The variable, d_w, denotes the nearest distance from any field node of interest to the wall surface. The most supporting variables occurring in previous formulations are estimated by

$$\begin{cases} f_{v3} = \dfrac{(1+\chi f_{v1})(1-f_{v2})}{\max(\chi, 0.001)} \\ f_{v2} = \left(1 + \dfrac{\chi}{C_{v2}}\right)^{-3} \\ S = \sqrt{2\Omega_{ij}\Omega_{ij}} \\ g = r + C_{w2}(r^6 - r) \end{cases} \quad (4.19)$$

where

$$\chi = \frac{\tilde{v}}{v_l}, \quad \Omega_{ij} = \frac{1}{2}\left(\frac{\partial v_i}{\partial x_j} - \frac{\partial v_j}{\partial x_i}\right), \quad r = \frac{\tilde{v}}{\tilde{S}\kappa^2 d_w^2} \quad (4.20)$$

The turbulent dynamic viscosity is computed using

$$\mu_T = f_{v1}\rho\tilde{v} \quad (4.21)$$

with

$$f_{v1} = \frac{\chi^3}{\chi^3 + C_{v1}^3} \tag{4.22}$$

All the constants used in the Spalart–Allmaras turbulence model are listed below:

$$\begin{cases} C_{b1} = 0.1355, C_{b2} = 0.622 \\ C_{v1} = 7.1, C_{v2} = 5 \\ \sigma = 2/3, \kappa = 0.41 \\ C_{w1} = C_{b1}/\kappa^2 + (1 + C_{b2})/\sigma, C_{w2} = 0.3, C_{w3} = 2 \\ C_{t1} = 1, C_{t2} = 2, C_{t3} = 1.3, C_{t4} = 0.5 \end{cases} \tag{4.23}$$

With the help of the Boussinesq eddy viscosity assumption [3], the effective dynamic viscosity (μ_{eff}) and the effective thermal conductivity (k_{eff}) are estimated as

$$\mu_{\text{eff}} = \mu + \mu_T, k_{\text{eff}} = k + k_T = c_p \left(\frac{\mu}{Pr} + \frac{\mu_T}{Pr_T} \right) \tag{4.24}$$

where μ, μ_T, k, and k_T are the laminar dynamic viscosity, turbulent dynamic viscosity, laminar thermal conductivity, and turbulent thermal conductivity, respectively. Pr and Pr_T denote the laminar and turbulent Prandtl numbers, respectively. For ideal air, $Pr = 0.72$ and $Pr_T = 0.9$. These effective quantities are used in the N–S equations for turbulent flows.

4.2 GSM for compressible Navier–Stokes equations

4.2.1 Discretization of governing equations

The PDEs that govern compressible flows, as shown in Eqs. (4.1)–(4.3) and other supporting PDEs, are discretized with GSM schemes to solve them numerically. These GSM schemes for gradient approximation have been discussed in detail and have been found to work well in Chapter 2. We are now using the preferable GSM scheme VII considering its excellent performance in stability and accuracy. Note that the 2nd-order ROE upwind scheme [4, 5] is necessarily employed for approximating the convective flux at the midpoint of the cell edge to ensure a desirable numerical stability for large Reynolds number problems [6, 7].

4.2.2 *Boundary conditions*

For a smoothing domain on the problem boundary, parts of the problem boundary are used to form the smoothing domains. At a corner point on the problem boundary, two or more boundary segments of the smoothing domain meet with each other. In such a case, the boundary node may be used in several boundary conditions.

- *Solid wall boundary condition*

For *inviscid* flows, slip wall conditions need to be imposed on the solid boundaries of the problem domain. When the viscous friction is ignored, the velocity vector must be tangential to the boundary surface. This implies that the normal velocity is zero. Hence, the contravariant velocity is zero. The convective flux terms related to velocity in the governing equations (4.1)–(4.3) vanish, and the pressure is explicitly specified as the pressure on the wall, p_w.

For *viscous* flows, non-slip wall conditions should be used. On the wall, $u_w = v_w = 0$, thus the momentum equations on the wall are not needed. The convective flux terms on the wall have the same forms as those for inviscid flows. In many cases, including this study, the wall is assumed to be adiabatic. Hence, it is not necessary to compute convective or viscous fluxes on the wall. The residuals of the momentum equations at the wall nodes should be set to zero, to prevent the generation of non-zero velocity components.

For high accuracy and consistency in prediction, the pressure on the wall used to compute the convective fluxes is [8]

$$p_w = \frac{1}{6}(5p_i + p_{j_k}) \tag{4.25}$$

This is also true for predicting the gradient on boundaries.

- *Far-field boundary condition*

The far-field conditions are imposed on the external bounds of the computational domains truncated from the infinite domain. The numerical implementation of the far-field conditions has to fulfill two basic requirements [9]. First, the truncation of the domain should be sufficiently large, so that it has no notable effects on the flow solution compared to the original infinite domain. Second, any outgoing perturbations must not be reflected back into the flow field. The far-field conditions based on the concept of characteristic variables are employed in this study for subsonic or transonic flows using GSM. This is done for both inflow and outflow.

For a subsonic inflow shown in Figure 4.1(a), the following formulations are imposed at the far-field bounds:

$$
\begin{cases}
p_b = \dfrac{1}{2}\{p_a + p_d - \rho_0 c_0 [n_x(u_a - u_d) + n_y(v_a - v_d)]\} \\[2mm]
\rho_b = \rho_a + \dfrac{p_a - p_d}{c_0^2} \\[2mm]
u_b = u_a - n_x \dfrac{p_a - p_b}{\rho_0 c_0} \\[2mm]
v_b = v_a - n_y \dfrac{p_a - p_b}{\rho_0 c_0}
\end{cases}
\tag{4.26}
$$

where ρ_0 and c_0 correspond to values at the reference state set for an interior point. The values at point a are determined from the freestream state.

For a subsonic outflow shown in Figure 4.1(b), the variables at far-field boundaries are set as

$$
\begin{cases}
p_b = p_a \\[2mm]
\rho_b = \rho_d + \dfrac{p_b - p_d}{c_0^2} \\[2mm]
u_b = u_d + n_x \dfrac{p_d - p_b}{\rho_0 c_0} \\[2mm]
v_b = v_a + n_y \dfrac{p_d - p_b}{\rho_0 c_0}
\end{cases}
\tag{4.27}
$$

with p_a being the prescribed static pressure.

In this study, one layer of dummy cells is generated on the far-field boundaries. It is used to determine the values of field variables and their gradients on the boundaries.

To shrink the computational domain, vortex correction [10] is also included in the computations on the far-field boundaries.

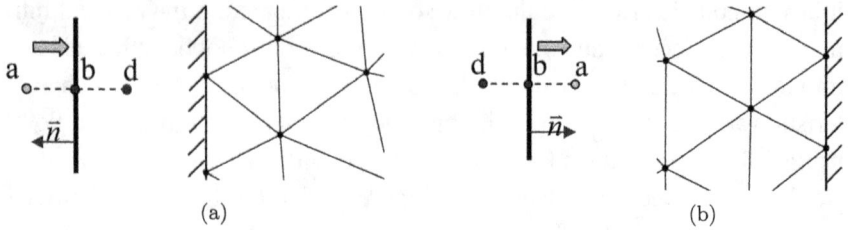

(a)   (b)

Figure 4.1.   Definition of far-field boundaries: (a) subsonic inflow; (b) subsonic outflow (Reprinted with permission from [5]).

- *Symmetric boundary condition*

At symmetrical boundaries, the following conditions are set:

$$\mathbf{n} \cdot \nabla U = 0$$

$$\mathbf{n} \cdot \nabla (\mathbf{t} \cdot \mathbf{v}) = 0$$

$$\mathbf{t} \cdot \nabla (\mathbf{n} \cdot \mathbf{v}) = 0 \tag{4.28}$$

where U denotes a scalar function of field variables, $\mathbf{n}$ and $\mathbf{t}$ represent the unit normal and tangential vectors, respectively, and $\mathbf{v}$ denotes the velocity vector. These conditions ensure vanishing normal component of velocity gradient, normal component of the gradient of tangential velocity, and tangential component of the gradient of normal velocity on symmetric boundary.

4.2.3   *Time integration of GSM*

In order to obtain numerical solutions for time-dependent transient flows, a time-marching scheme is necessarily combined with spatial discretization. Specifically, a dual time-stepping approach [5] is employed which involves adding a pseudo-temporal term to the physical temporal term. This enables the computation to march along the user-specified physical time (Δt) at each pseudo-time step.

During each physical time step, we pursue steady-state solutions with respect to a pseudo-time step ($\Delta \tau$) through a number of iterations. Once the steady-state field is obtained, we proceed to the next pseudo-time step. Compared to a single global time-stepping approach, the dual time-stepping approach allows for larger physical time steps to be taken, which can lead to faster convergence and improved stability. In cases where only a steady-state solution is needed, a single iteration in the physical time is sufficient. The implementation detail of the dual time-stepping approach is given as follows.

Take the Eq. (4.1) as an example. First, the PDE is rewritten in the following residual form with respect to the pseudo-time (τ):

$$\frac{\partial \rho}{\partial \tau} + \frac{\partial \rho}{\partial t} + \nabla \cdot (\rho \mathbf{v}) = 0, \quad \text{i.e.} \quad \frac{\partial \rho}{\partial \tau} = -\left(\frac{\partial \rho}{\partial t} + \nabla \cdot (\rho \mathbf{v})\right) \quad (4.29)$$

For the solution at $(n+1)^{th}$ physical time and the $(m+1)^{th}$ pseudo-time, the discretized equations can be written as

$$\frac{\Delta \rho^m}{\Delta \tau} = -\left(\frac{\partial \rho^n}{\partial t} + \nabla \cdot (\rho \mathbf{v})^n\right) \quad (4.30)$$

In Eq. (4.30), $\Delta \rho^m$ denotes the difference of solutions between the two consecutive pseudo-times, which is calculated as $\Delta \rho^m = \Delta \rho^{m+1} - \Delta \rho^m$. In simulations, the 2nd-order 3-level backward differencing scheme is used to approximate the physical temporal terms

$$\frac{\partial \rho^n}{\partial t} \approx \frac{3\rho^{n+1} - 2\rho^n + \rho^{n-1}}{2\Delta t} \quad (4.31)$$

This formulation is explicit, meaning that the solution at $(n+1)$-th step can be determined based on the known information at n-th step and $(n-1)$-th step, without the need of solving a system of equations. Hence, it is stable under the condition that the time step is sufficiently small, known as the Courant–Friedrichs–Lewy (CFL) condition [11]. Because this scheme involves solutions at the previous two time levels, it is known as A-stable [12], which allows one to use a large physical time step in the iterations.

For marching in pseudo-time, a number of schemes can be taken for GSM, including the explicit multistage Runge–Kutta (RK) method [12–14], the point implicit multistage RK method [15], and the implicit Lower Upper Symmetrical Gauss–Seidel (LUSGS) method [16–18]. More details can be found in [5].

4.3 Numerical examples

This section presents GSM applications for conservative, compressible flows. Cases studied in this section involve the following:

- *Case I*: Steady-state inviscid flow over the NACA0012 airfoil. A GSM Euler solver is used for this case.
- *Case II*: Laminar flow over a flat plate. A GSM N-S solver is used.
- *Case III*: Turbulent flow over the RAE2822 airfoil. A GSM N-S solver with a turbulence model is used for this case.
- *Case IV*: Transient flow over the circular cylinder. A GSM N-S solver with a turbulence model is used.

4.3.1 *Case I: Inviscid flow over the NACA0012 airfoil*

The GSM Euler solver using multiple GSDs is applied to solve an inviscid flow over the NACA0012 airfoil, using unstructured triangular grids. The setting of the problem is as follows:

- For the freestream, $T_\infty = 288$K, $p_\infty = 1.0 \times 10^5$ Pa, $Ma = 0.8$.
- The angle of attack $\alpha = 1.25$.

where T_∞, p_∞, and Ma denote, respectively, the temperature, static pressure, and Mach number of the freestream. Numerical results are presented below.

Figure 4.2 shows the unstructured grid used in the GSM computation. Contours of the density, the static pressure, and Mach number are plotted in Figure 4.3. It is seen that the strong shock occurring on the upper surface of the airfoil is clearly captured, as well as the weaker shock on the lower surface. The results are qualitatively agreeable with published results [13].

A comparison between the implicit matrix-free LUSGS method and the explicit RK5 method for marching the pseudo-time in GSM is conducted. It is a tedious for end user to seek a suitable CFL number for rapid convergent solutions. We thus present our finding for future reference. The improvement in computational efficiency in the LUSGS method can be seen clearly in Figure 4.4 and Figure 4.6, in comparison with the explicit RK5 method.

(a) Boundary conditions

(b) Computational grid used in this testing case

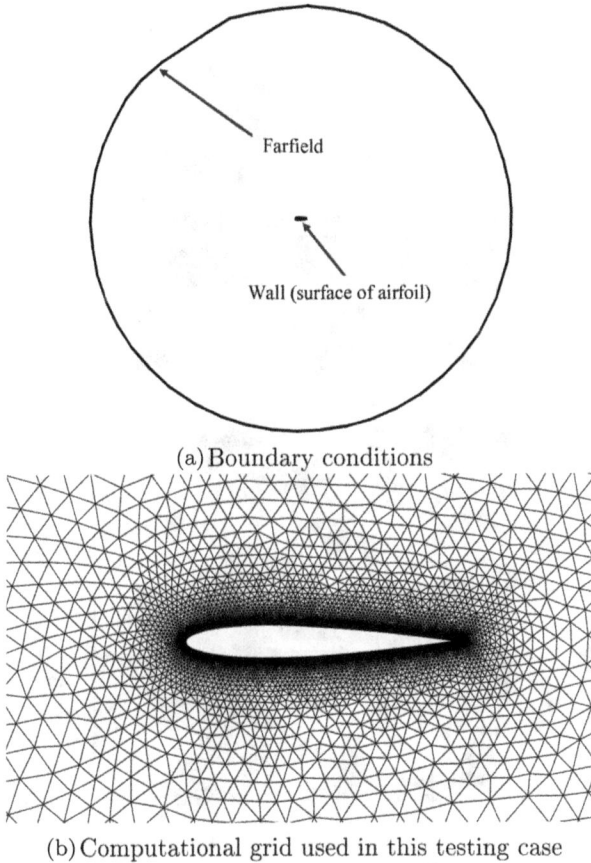

Figure 4.2. Boundary conditions and irregular grid used in the GSM simulation of the flow over the NACA0012 airfoil (Reprinted with permission from [5]).

Additional sets of grids are also created for this study. Figure 4.4 shows the convergence curves for these two methods using two sets of grids. Apparently, to achieve the same convergence, the LUSGS method does not save the number of iterations, as compared to the explicit RK5 method. However, the convergence curves of the LUSGS method seem more stable than those of the explicit RK5. This may be due to the fact that the LUSGS is more tolerant to the CFL number. In the convergence curves of numerical error *vs* the CPU time, the advantages of the matrix-free LUSGS method

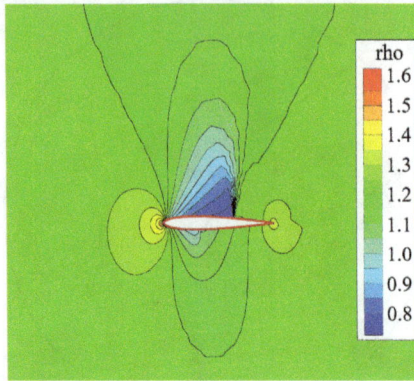

(a) Density

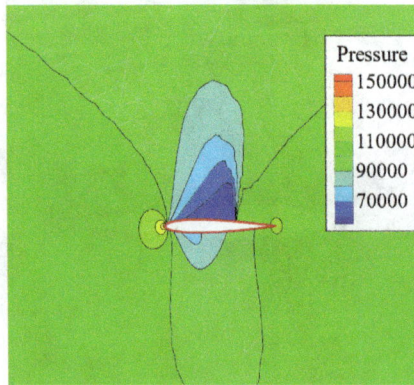

(b) Static pressure

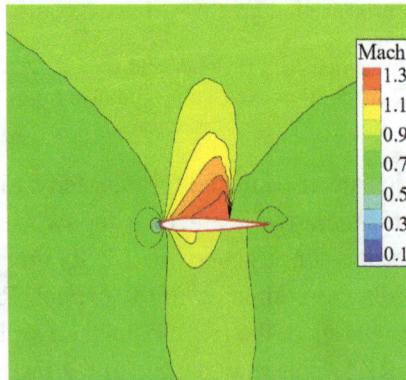

(c) Mach number

Figure 4.3. Spatial distribution of variables across the field in the flow over the NACA0012 airfoil obtained using GSM Euler solver (Reprinted with permission from [19]).

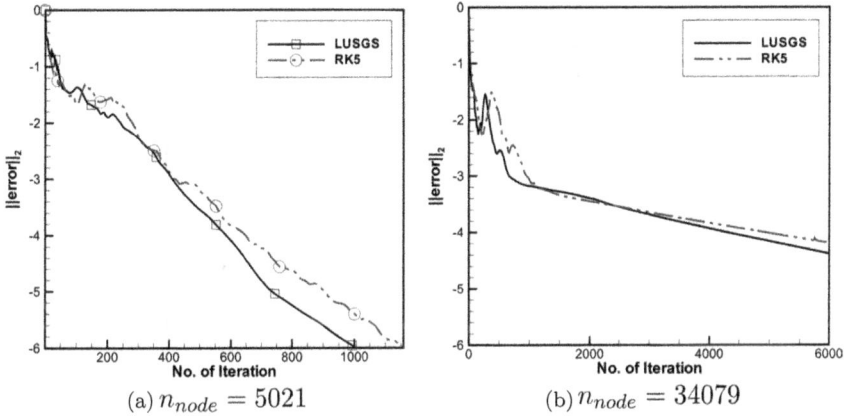

(a) $n_{node} = 5021$ (b) $n_{node} = 34079$

Figure 4.4. Comparison between the matrix-free LUSGS and RK5 methods implemented in GSM in term of convergence curve (residual vs iteration) for solutions to the flow over the NACA0012 airfoil (Reprinted with permission from [5]).

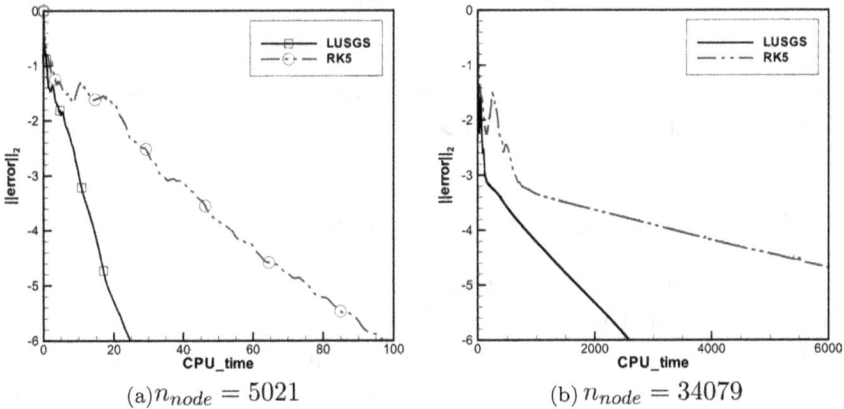

(a) $n_{node} = 5021$ (b) $n_{node} = 34079$

Figure 4.5. Comparison between the matrix-free LUSGS and RK5 methods implemented in GSM in terms of computational efficiency for solutions to the flow over the NACA0012 airfoil (Reprinted with permission from [5]).

over the explicit RK5 method become obvious, as shown in Figure 4.5. For both sets of grids, the LUSGS method leads to greater savings in the computational time, compared with the RK5. As the number of nodes increases, the LUSGS becomes much more cost effective than the RK5, as

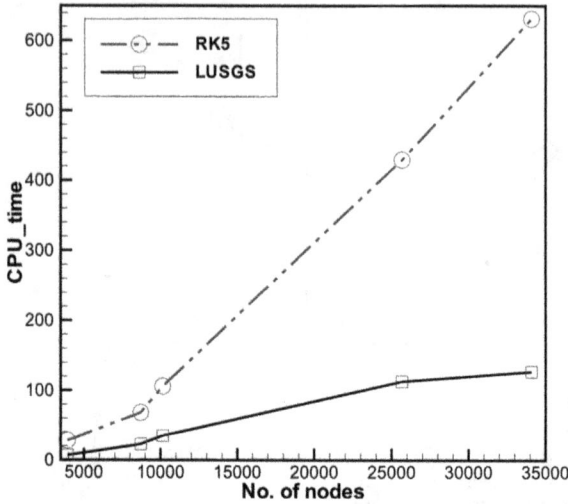

Figure 4.6. Comparison between the matrix-free LUSGS and the RK5 methods in terms of required CPU time vs averaged nodal spacing in the GSM solutions to the flow over the NACA0012 airfoil (Reprinted with permission from [5]).

shown in Figure 4.6. The computational efficiency in the LUSGS method is achieved without compromise of computational accuracy.

4.3.2 *Case II: Laminar flow over the flat plate*

This testing case simulates the laminar flows over a flat plate, using the GSM N–S solver. The plate is treated as a rigid wall, the freestream is set with the conditions $Re = 5000$ and $Ma = 0.5$, and the angle of attack is set to *zero*. A rectangular computational domain with right triangular grids is generated, as depicted in Figure 4.7. Non-slip conditions are imposed onto the wall and symmetry conditions are applied to the region ahead of the plate along the x axis (horizontal). Far-field conditions are imposed at the external boundaries.

Figure 4.8 plots the vector field of the velocity, where boundary layer effect is clearly shown near the wall surface. The profile of the predicted skin friction coefficient C_f varying with the distance x is compared with the analytical Blasius solution, as shown in Figure 4.9. It is apparent that the GSM numerical results show good agreement with the analytical solution.

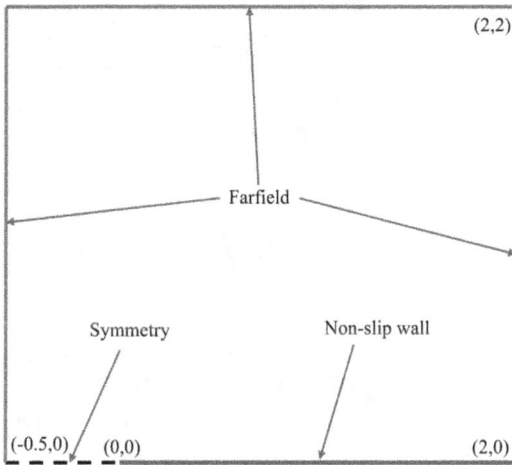

(a) Boundary conditions

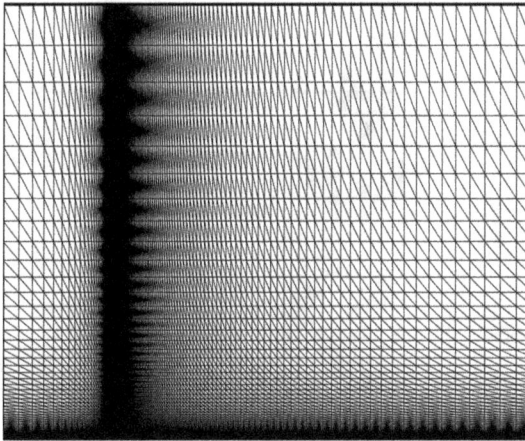

(b) Computational grid

Figure 4.7. Boundary conditions and computational grid used in GSM for a laminar flow over a flat rigid plate (Reprinted with permission from [5]).

4.3.3 Case III: Turbulent flow over the RAE2822 airfoil

In this testing case, the GSM N–S solver is used to simulate turbulent flows over the RAE2822 airfoil, using unstructured triangular grids, as depicted in Figure 4.10. The freestream is set with the following conditions: $T_\infty = 255.556\,\text{K}$, $p_\infty = 1.0756256 \times 10^5\,\text{Pa}$, $Re = 6.5 \times 10^6$, $Ma = 0.729$, and

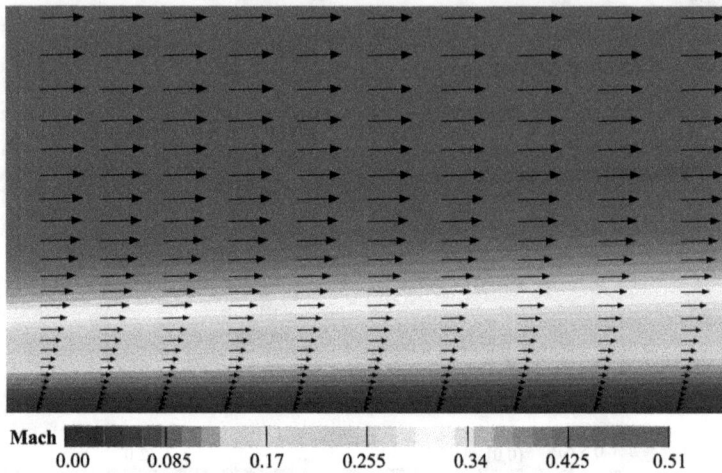

Figure 4.8. Velocity profiles near boundary with contour of Mach number coded in color (Reprinted with permission from [19]).

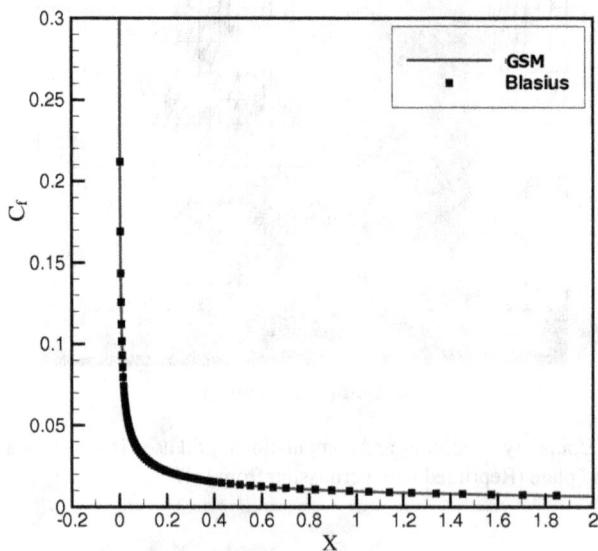

Figure 4.9. Comparison of the GSM predicated and analytical wall friction coefficient (Reprinted with permission from [19]).

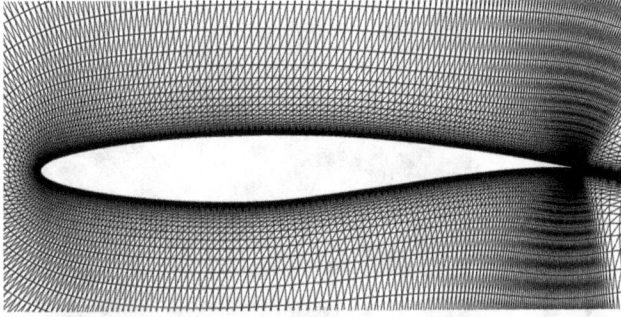

Figure 4.10. Viscous triangular cells around RAE2822 airfoil used in GSM (Reprinted with permission from [19]).

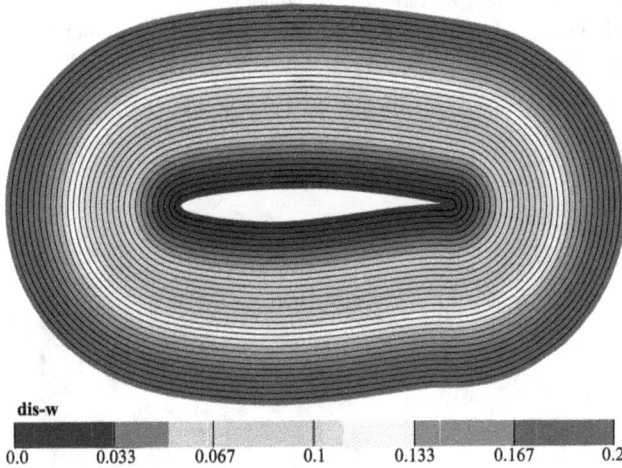

Figure 4.11. Spatial distribution of nearest wall distance around RAE2822 airfoil (Reprinted with permission from [5]).

$\alpha = 2.31°$. The Reynolds number Re is evaluated with airflow parameters for the freestream and the chord length of the airfoil.

Figure 4.11 plots the spatial distribution of the nearest wall distance of the nodes used in this turbulent flow problem. The spatially distributed turbulent dynamic viscosity is plotted in Figure 4.12. It is seen clearly that the turbulence originates from the wall and is magnified along the trailing edge of the airfoil. The contour of the Mach number is plotted in Figure 4.13.

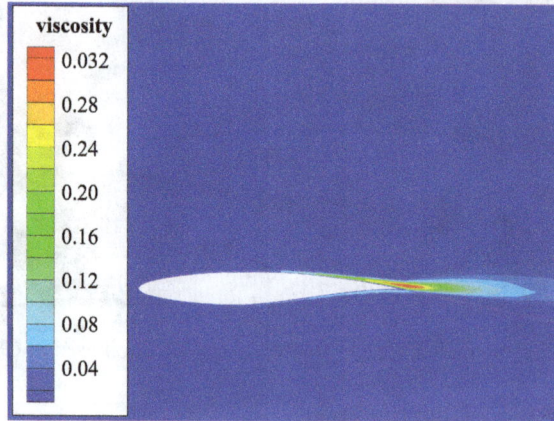

Figure 4.12. Contour plot of turbulent dynamic viscosity (μ_T) in the flow over the RAE2822 airfoil obtained using GSM solver (Reprinted with permission from [5]).

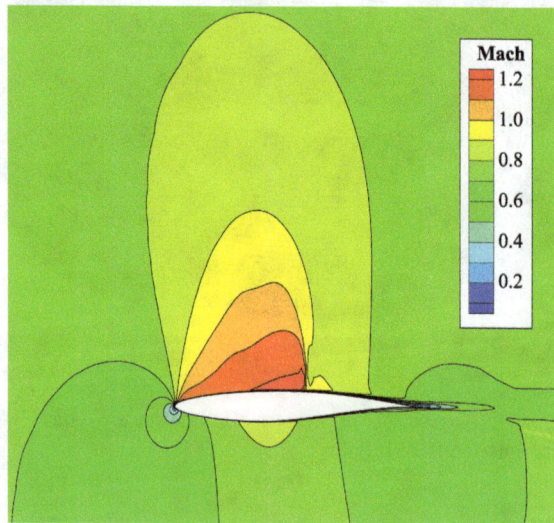

Figure 4.13. Contour plot of Mach number in the flow over the RAE2822 airfoil obtained using GSM solver (Reprinted with permission from [19]).

It is seen that a shock occurring on the upper side of the airfoil is captured. The GSM predicted pressure coefficient (C_p) distributed on the airfoil surface is plotted in Figure 4.14, in comparison with the results from the commercial CFD package FLUENT (FVM-based) using the same grid

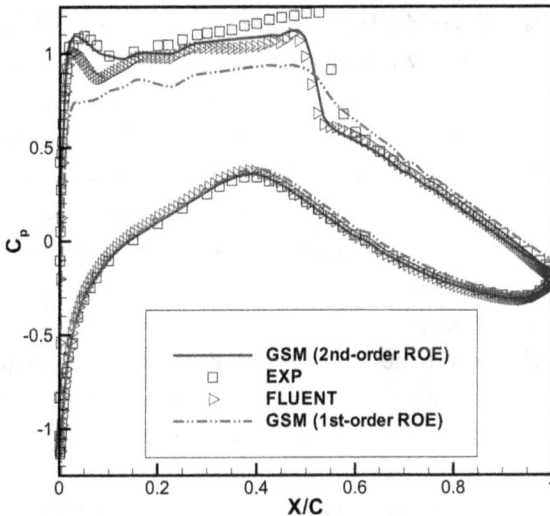

Figure 4.14. Profiles of pressure coefficients on the surface of an RAE2822 airfoil obtained using GSM in comparison with FLUENT and experimental data (Reprinted with permission from [5]).

and experimental data [20]. It is found that the GSM result is more accurate than the FLUENT result, especially near the leading edge. There is noticeable difference between our GSM solution and experimental data occurring around the shock region. This is partially due to the insufficiently fine cells used across the shock. This can be further improved if one uses the GSM with solution-based adaptive techniques [21–23].

4.3.4 Case IV: Time-dependent compressible laminar flow over a circular cylinder

Flows over a circular cylinder are a classic benchmarking problem and are widely used to verify the effectiveness of numerical methods. The problem may look simple, but it is not trivial to get the correct behavior via numerical means. It is thus also used in this study to demonstrate the reliability of the GSM N–S solver applied to solving time-dependent problems using the dual time-stepping approach.

Consider a circular cylinder with the diameter $D = 1.0$ m. The freestream is set under the following conditions: $T_\infty = 277.778$K,

$p_\infty = 3.1127$ Pa, $u_\infty = 66.8163$ m/s, $Re = 150, Ma = 0.2$, and $\alpha = 0°$. Under these conditions, vortex shedding is expected to occur in the wake flow behind the cylinder and should be correctly predicted.

The GSM N-S solver employs the implicit LUSGS method to handle the pseudo-time domain. The computational domain discretized with a triangular grid is used in this study, as shown in Figures 4.15 and 4.16. To resolve the flow features accurately, finer cells are used near the cylinder surface and within the wake regions, where larger gradients and complex flow behavior may take place.

To accurately capture the behavior of vortex shedding, the physical time step (Δt) used in the dual time approach is set as low as 1.0×10^{-4}. At each physical time step, the iteration terminates when the convergence error satisfies $\varepsilon_{con} \le 1.0 \times 10^{-4}$.

The evolutions of the vorticity and Mach number within a typical shedding period are computed and presented in Figures 4.17 and 4.18, respectively. The time evolutions of predicted drag and lift coefficients on the cylinder surface are plotted in Figure 4.19, with respect to the number of physical time iterations (Figure 4.19(a)), and the physical flow time (Figure 4.19(b)). It is observed that the transition period takes about 1 second, and the flow becomes stabilized, giving rise to periodic oscillations of drag and lift coefficients with respect to physical time.

The period for the vortex shedding is predicted from the periodic change of lift coefficient in the stabilized stage. The drag coefficient is not

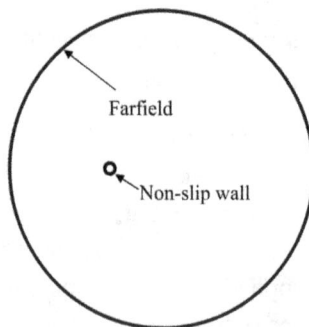

Figure 4.15. Geometrical and boundary conditions for compressible flow over the circular cylinder with non-slip wall condition (Reprinted with permission from [5]).

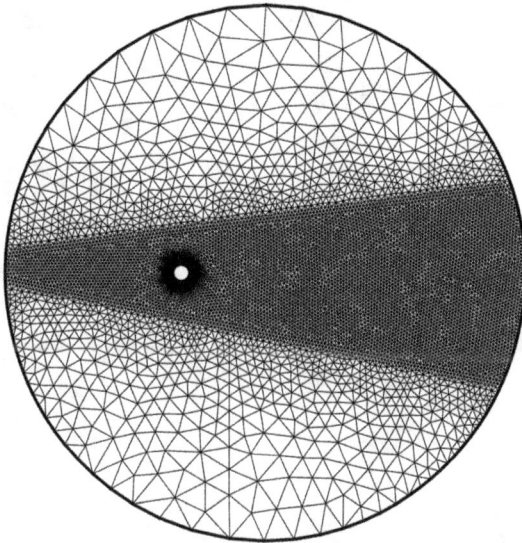

(a) overall view

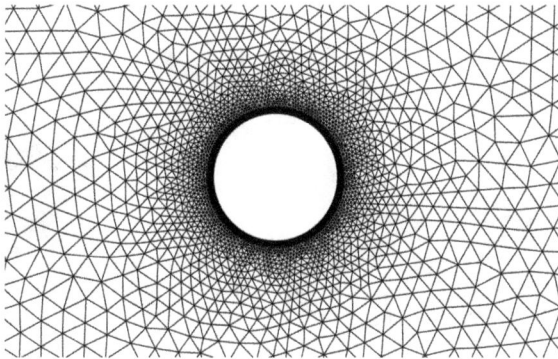

(b) close-up view

Figure 4.16. Triangular grid generated for the GSM simulation of compressible flow over the circular cylinder ($n_{node} = 10079$, $n_e = 19696$) (Reprinted with permission from [5]).

recommended to be used in the determination of the shedding period, since the drag coefficient varies from the vortex shedding from both sides of the cylinder. Because of this, the periodic change of the drag coefficient is half of that obtained using the lift coefficient. In this study, using Figure 4.19(b), the period T_p for the vortex shedding is found to be about 0.083 second.

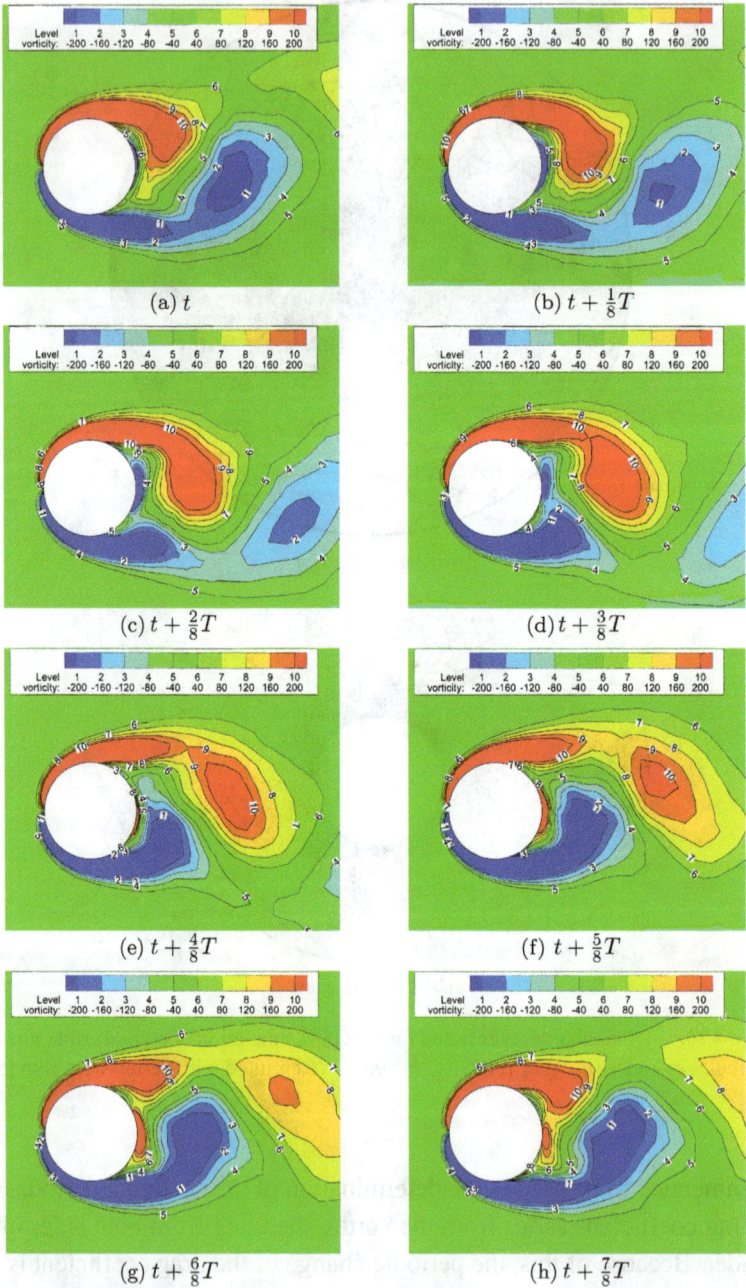

Figure 4.17. Contour of vorticity within a stabilized vortex-shedding cycle in the vicinity of the cylinder obtained using GSM solver (Reprinted with permission from [5]).

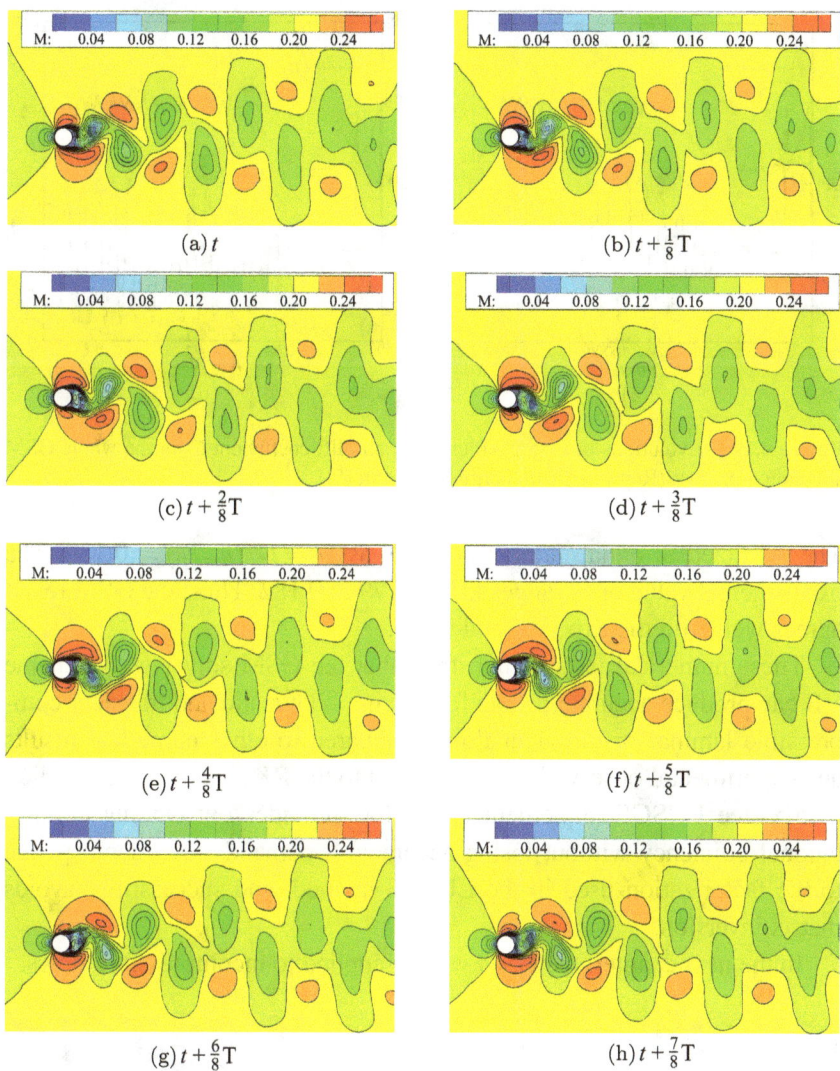

Figure 4.18. Contour of Mach number within a stabilized vortex-shedding cycle obtained using GSM solver (Reprinted with permission from [5]).

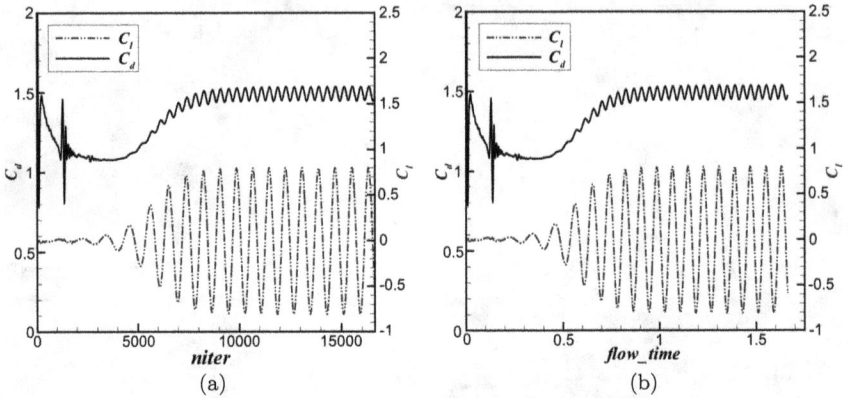

Figure 4.19. Evolutions of drag coefficient C_d and lift coefficient C_l in the whole GSM simulation process (Reprinted with permission from [5]).

The predicted Strouhal number is found as $St = D/(u_\infty T_p) = 0.180$. The experimental value of St ranges from 0.179 to 0.182. Thus, the GSM result agrees well with the experimental value.

These numerical tests demonstrate that the GSM solver with Scheme VII can produce accurate results for solving both static and transient compressible laminar or turbulent flows compared to other numerical results and experimental observations. Compared to the RK5 method, the implicit matrix-free LUSGS implemented in GSM provides outstanding computational efficiency without losing accuracy. It demands less memory, due to the factorization used in the LUSGS method, and successfully avoids matrix inversion.

In the next chapter, the GSM will be applied to incompressible flows.

References

[1] William Sutherland, The viscosity of gases and molecular force, *Philosophical Magazine*, 5(36), 507–531 (1893).

[2] P. R. Spalart and S. R. Allmaras, One-equation turbulence model for aerodynamic flows, *Recherche aerospatiale*, 1(1), 5–21, (1994). doi: 10.2514/6.1992-439.

[3] F. G. Schmitt, About Boussinesq's turbulent viscosity hypothesis: historical remarks and a direct evaluation of its validity, *Comptes Rendus Mécanique*, 335(9–10), 617–627 (2007, September). doi: 10.1016/J.CRME.2007.08.004.

[4] P. L. Roe, Approximate Riemann solvers, parameter vectors, and difference schemes, *Journal of Computational Physics*, 43(2), 357–372 (1981, October 01). doi: 10.1016/0021-9991(81)90128-5.

[5] X. Xu, Development of Gradient Smoothing Method (GSM) for fluid flow problems, National University of Singapore, (2009). doi: https://scholarbank.nus.edu.sg/handle/10635/160959.

[6] J. Blazek, *Computational Fluid Dynamics: Principles and Applications: Third Edition.* Elsevier Ltd, 2015. doi: 10.1016/C2013-0-19038-1.

[7] G. R. Liu and Y. T. Gu, *An Introduction to Meshfree Methods and Their Programming,* 1st ed. Netherlands: Springer Netherlands, 2005. Accessed: November 09, 2019. Available: https://www.springer.com/gp/book/9781402032288.

[8] H. Luo, J. D. Baum, and R. Löhner, An improved finite volume scheme for compressible flows on unstructured grids. *33rd Aerospace Sciences Meeting and Exhibit,* American Institute of Aeronautics and Astronautics Inc, AIAA, (1995). doi: 10.2514/6.1995-348.

[9] J. Blazek, *Computational Fluid Dynamics: Principles and Applications: Third Edition.* Elsevier Ltd, (2015). doi: 10.1016/C2013-0-19038-1.

[10] W. J. Usab, *Embedded Mesh Solutions of the Euler Equation using a Multiple-Grid Method,* Massachusetts Institute of Technology (1983).

[11] R.; F. K.; L. H. Courant, On the partial difference equations of mathematical physics, *IBM Journal of Research and Development,* 11(2), 215–234 (1967).

[12] V. Venkatakrishnan, Convergence to Steady State Solutions of the Euler Equations on Unstructured Grids with Limiters, *Journal of Computational Physics,* 118(1), 120–130, (1995, April). doi: 10.1006/jcph.1995.1084.

[13] T. J. Barth, Numerical methods for conservation laws on structured and unstructured meshes, Technical report, VKI Lecture Series 2003 (2003).

[14] A. Arnone, M. S. Liou, and L. A. Povinelli, Multigrid time-accurate integration of navier-stokes equations, in *11th Computational Fluid Dynamics Conference, 1993,* American Institute of Aeronautics and Astronautics Inc, AIAA, 1993, 694–702. doi: 10.2514/6.1993-3361.

[15] N. D. Melson and H. L. Atkins, Time-accurate Navier-Stokes calculations with multigrid acceleration. In *6th Copper Mountain Conference on Multigrid Methods* (1993, November).

[16] A. Jameson and E. Türkei, Implicit Schemes and LU Decompositions* (1981).

[17] H. Rieger and A. Jameson, Solution of steady three-dimensional compressible Euler and Navier-Stokes equations by an implicit LU scheme. AIAA paper, American Institute of Aeronautics and Astronautics (AIAA), (1988, January). doi: 10.2514/6.1988-619.

[18] S. Yoon and A. Jameson, Lower-upper symmetric-gauss-seidel method for the euler and navier-stokes equations, *AIAA Journal,* 26(9), 1025–1026 (1988, May). doi: 10.2514/3.10007.

[19] G. R. Liu and G. X. Xu, A gradient smoothing method (GSM) for fluid dynamics problems, *International Journal for Numerical Methods in Fluids,* 58(10), 1101–1133 (2008, December).

[20] P. Cook, M. McDonald, and M. Firmin, Aerofoil RAE 2822: Pressure distributions, and boundary layer and measurements. Technical Report AGARD Report AR 138, (1979).

[21] J. Zhang, G. R. Liu, K. Y. Lam, H. Li, and G. Xu, A gradient smoothing method (GSM) based on strong form governing equation for adaptive analysis of solid

mechanics problems, *Finite Elements in Analysis and Design*, 44(15), 889–909 (2008, November).

[22] J. Yao, T. Lin, G. R. Liu, and C. L. Chen, An adaptive GSM-CFD solver and its application to shock-wave boundary layer interaction, *International Journal of Numerical Methods for Heat & Fluid*, 25(6), 1282–1310 (2015).

[23] G. X. Xu, G. R. Liu, and A. Tani, "An adaptive gradient smoothing method (GSM) for fluid dynamics problems," *International Journal for Numerical Methods in Fluids*, 62(5), 499–529 (2010, February). doi: 10.1002/fld.2032.

Chapter 5

Eulerian GSM for Incompressible Flows

Contents

This chapter applies GSM to a series of incompressible flows. The presentation is given in the following order:

- Section 5.1 introduces the governing equations of incompressible flows that will be solved with GSM.
- Section 5.2 presents the necessary numerical treatments in GSM to solve the governing PDEs.
- Section 5.3 shows the GSM solutions for incompressible flows, including steady-state flow passing a circular cylinder, steady-state flow over

a backstep, steady-state lid-driven cavity flow, and time-dependent flow over a cylinder.

- Section 5.4 summarizes the performance of GSM in comparison with the widely used FDM.

5.1 Setting of the problem

5.1.1 *Governing equations for incompressible flows*

GSM works naturally for incompressible flows governed by the incompressible Navier–Stokes equation. In this study, we will ignore the heat transfer and consider only the continuity and momentum equations in 2D cases. The continuity equation is

$$\frac{\partial u}{\partial x} + \frac{\partial v}{\partial y} = 0 \tag{5.1}$$

The momentum equations are

$$\frac{\partial \rho u}{\partial t} + u\frac{\partial \rho u}{\partial x} + v\frac{\partial \rho u}{\partial y} + \frac{\partial p}{\partial x} = \frac{\partial \tau_{xx}}{\partial x} + \frac{\partial \tau_{xy}}{\partial y} \tag{5.2}$$

$$\frac{\partial \rho v}{\partial t} + u\frac{\partial \rho v}{\partial x} + v\frac{\partial \rho v}{\partial y} + \frac{\partial p}{\partial y} = \frac{\partial \tau_{xy}}{\partial x} + \frac{\partial \tau_{yy}}{\partial y} \tag{5.3}$$

The set equations can be rewritten in the following conservative form:

$$\frac{\partial \rho u}{\partial x} + \frac{\partial \rho v}{\partial y} = 0 \tag{5.4}$$

$$\frac{\partial \rho u}{\partial t} + \frac{\partial(\rho u^2 + p)}{\partial x} + \frac{\partial \rho uv}{\partial y} = \frac{\partial \tau_{xx}}{\partial x} + \frac{\partial \tau_{xy}}{\partial y} \tag{5.5}$$

$$\frac{\partial \rho v}{\partial t} + \frac{\partial \rho uv}{\partial x} + \frac{\partial(\rho v^2 + p)}{\partial y} = \frac{\partial \tau_{xy}}{\partial x} + \frac{\partial \tau_{yy}}{\partial y} \tag{5.6}$$

Variables ρ, p, u, and v denote, respectively, the density, static pressure, and velocity components in x and y directions.

For incompressible non-Newtonian flows, with the help of Stokes' theorem, viscous stresses become

$$
\begin{cases}
\tau_{xx} = 2\mu \dfrac{\partial u}{\partial x} \\[2mm]
\tau_{yy} = 2\mu \dfrac{\partial v}{\partial y} \\[2mm]
\tau_{xy} = \tau_{yx} = \mu \left(\dfrac{\partial u}{\partial y} + \dfrac{\partial v}{\partial x} \right)
\end{cases}
\tag{5.7}
$$

where μ denotes the dynamical viscosity.

Because the flow is incompressible, the density ρ stays constant. Hence, Eqs. (5.4)–(5.6) are already enclosed. We thus have a total of three equations and three unknowns: pressure (p) and two velocity components (u and v).

Note that there is no temporal term in the continuity equation. Therefore, a temporal term in pseudo-time is added into these equations, together with artificial compressibility [1]. This allows us to treat incompressible flows in a similar way as the compressible flows. The augmented PDEs become

$$
\frac{1}{\beta} \frac{\partial p}{\partial \tau} + \frac{\partial \rho u}{\partial x} + \frac{\partial \rho v}{\partial y} = 0
\tag{5.8}
$$

$$
\frac{\partial \rho u}{\partial \tau} + \frac{\partial \rho u}{\partial t} + \frac{\partial (\rho u^2 + p)}{\partial x} + \frac{\partial \rho u v}{\partial y} = \frac{\partial \tau_{xx}}{\partial x} + \frac{\partial \tau_{xy}}{\partial y}
\tag{5.9}
$$

$$
\frac{\partial \rho v}{\partial \tau} + \frac{\partial \rho v}{\partial t} + \frac{\partial \rho u v}{\partial x} + \frac{\partial (\rho x^2 + p)}{\partial y} = \frac{\partial \tau_{xy}}{\partial x} + \frac{\partial \tau_{yy}}{\partial y}
\tag{5.10}
$$

The additional parameter, β, is called *artificial compressibility*. A dimensional analysis finds β having the same unit as velocity, i.e., in m/s. The square root of the artificial compressibility β can be interpreted as the speed of artificial pressure wave. In practice, the value of β is set with a prescribed value during simulation. Since it can significantly affect the convergence of the iterative solution procedure, care is needed in choosing β.

5.2 GSM for incompressible flows

5.2.1 *Discretization of governing equations*

Similar to the compressible flows in Chapter 4, the viscous terms in the governing PDEs are approximated with the GSM scheme VII while the convection terms are evaluated with the 2nd-order ROE flux-difference splitting upwind scheme [2–5] when applied to high-Reynolds number problems.

5.2.2 *Boundary conditions*

Non-slip viscous wall boundary conditions and characteristic variable boundary conditions for freestream are adopted in this study.

On physical walls, velocity components are set to be zero accordingly. Thus, the two momentum equations, (5.9) and (5.10), are not solved but the modified continuity equation (5.8) with respect to pressure (p) is required to be solved in order to approximate the pressure on the wall.

The characteristic far-field conditions are derived upon the inviscid governing equations and vary with the first eigenvalue (λ_1), which is equal to the contravariant velocity. For inflowing boundaries, the following inflow conditions are used at the boundary:

$$
\begin{cases}
p_b = \dfrac{p_\infty + p_a}{2} - \dfrac{1}{2c_0}[\beta(V_\infty - V_a) - V_0(p_\infty - p_a)] \\[2mm]
u_b = u_\infty - \dfrac{p_\infty - p_b}{\beta c_0}[u_0(V_0 + c_0) + \beta n_x] \\[2mm]
v_b = v_\infty - \dfrac{p_\infty - p_b}{\beta c_0}[v_0(V_0 + c_0) + \beta n_y]
\end{cases}
\tag{5.11}
$$

For outflowing boundaries, the following outflow conditions are imposed:

$$
\begin{cases}
p_b = p_\infty \\[2mm]
u_b = u_a - \dfrac{p_a - p_b}{\beta c_0}[u_0(V_0 - c_0) + \beta n_x] \\[2mm]
v_b = v_a - \dfrac{p_a - p_b}{\beta c_0}[v_0(V_0 - c_0) + \beta n_y]
\end{cases}
\tag{5.12}
$$

In Eqs. (5.11) and (5.12), subscripts ∞, a, and b, respectively, denote the flow conditions at infinity, interior, and boundary nodes. Reference values for u_0, v_0, c_0, and V_0 are evaluated at interior nodes. Details about the derivation of characteristic variable boundary conditions can be found in [4].

It should be noted that in the incompressible GSM code, Eqs. (5.11) and (5.12) are used to update the unknowns at dummy nodes, which are positioned right at the boundary nodes. The unknowns at physical far-field boundary nodes are updated during Runge–Kutta iterative loops.

5.3 Numerical examples

The above-mentioned incompressible GSM solver has tested for solutions to some classical incompressible flow problems. Here, the attained steady-state numerical results for flows over a circular cylinder, over a backstep, and in a lid-driven cavity are first presented and discussed. Then, the results for time-dependent flow past a circular cylinder are presented.

5.3.1 *Steady-state flow past a circular cylinder*

The flow over a circular cylinder is a common benchmark problem for incompressible fluid flows. A circular cylinder of diameter $D = 1$ is placed in an incompressible viscous fluid with far-field conditions imposed at the outer boundary and non-slip wall conditions on the cylinder's surface, as shown in Figure 5.1. The set of grids which was adopted in the GSM

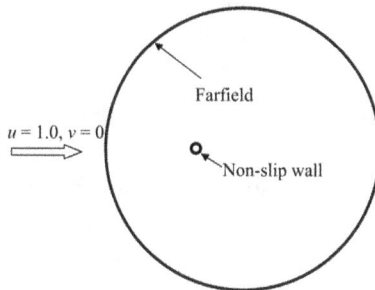

Figure 5.1. Geometrical and boundary conditions in the case of flow over a cylinder (Reprinted with permission from [10]).

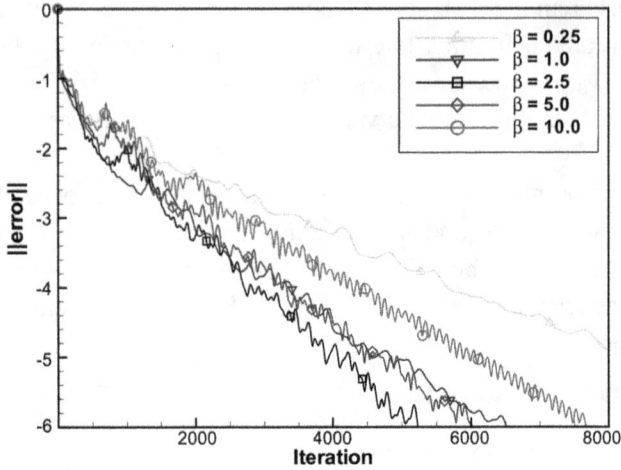

Figure 5.2. Effect of artificial compressibility on the convergence (Reprinted with permission from [10]).

solutions for compressible flow over the circular cylinder in Chapter 4 is reused in this case.

The choice of artificial compressibility value, β, can significantly affect convergence and stability, and an optimal value of β depends on the problem being solved. While Kwak *et al.* [6] suggested that an artificial compressibility value of β in the range of $0.1 \sim 10.0$ should work well for most problems, the numerical results presented in Figure 5.2 show that $\beta = 2.5$ yields the most rapid and efficient convergence for this problem.

The flow is characterized by the Reynolds number, Re; as Re increases beyond 40, the flow becomes unstable and forms periodic vortex shedding known as a Von Karman vortex street. The steady-state solutions will be illustrated in this section first and the time-dependent results will be presented in Section 5.3.4.

Figure 5.3 shows the formations of eddies behind the cylinder in term of streamlines, as the Reynolds number increases. It is seen that as the Reynolds number increases, the pair of symmetrical eddies behind the cylinder grows in scale, and the reattachment length (S) is measured from the right endpoint of the cylinder to the reattachment point where the two eddies end. The attained reattachment lengths normalized by the cylinder's

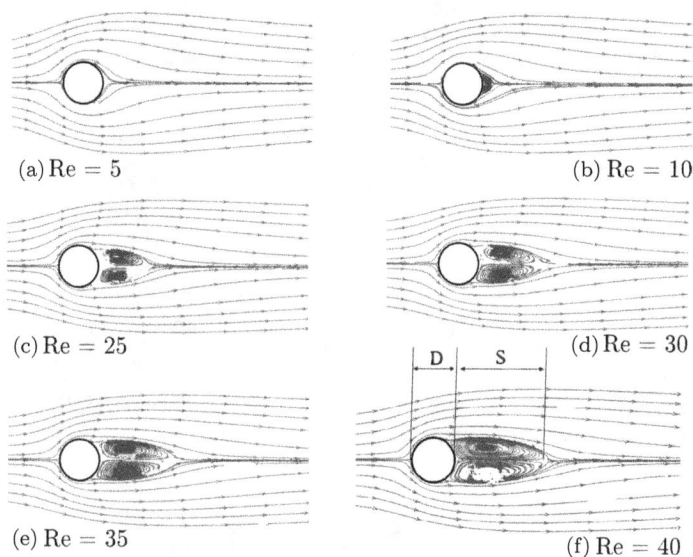

Figure 5.3. Plots of streamlines in the steady-state flows over a circular cylinder with various Reynolds numbers (Reprinted with permission from [10]).

Table 5.1. Comparison of predicted reattachment length ratio (S/D). The "Exp" result is from [7] and the "DDM" result is from [8].

Re	Exp	DDM	GSM
10	0.25	—	0.25
25	1.15	0.99	1.12
30	1.49	1.27	1.5
35	1.8	1.53	1.84
40	2.2	1.94	2.22
42	2.35	2.09	2.26

diameter are summarized and compared to experimental [7] and numerical results [8] in Table 5.1 and Figure 5.4, with good agreement to GSM solutions.

The profiles of drag coefficient, C_D, versus the Reynolds number are also analyzed in Figure 5.5, with C_D decreasing as the Reynolds number

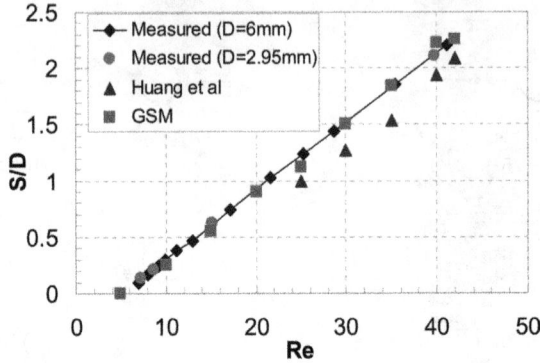

Figure 5.4. Comparison of computed reattachment length ratios (S/D) with varied Reynolds number (Reprinted with permission from [10]).

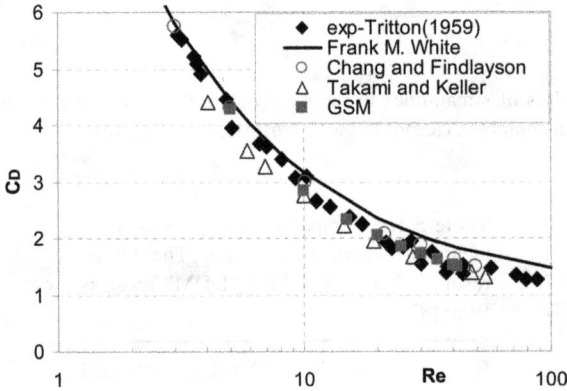

Figure 5.5. Comparison of computed drag coefficients on the cylinder surface with varied Reynolds number (Reprinted with permission from [10]).

increases within the laminar regime. The GSM solver accurately captures this trend and outperforms other numerical solutions in terms of accuracy when compared to the experimental data [9].

5.3.2 *Steady-state flow over a backstep*

The second benchmark case is about the steady-state flow over a sudden backstep. The incoming flow at the inlet is assumed to be fully developed, consistent with a parabolic distribution function. The fluid flows through a channel bounded by top and bottom walls and exits at the other end.

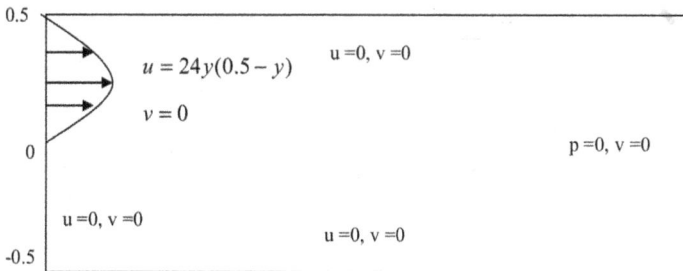

Figure 5.6. Geometrical and boundary conditions for the flow problem over a sudden backstep (Reprinted with permission from [10]).

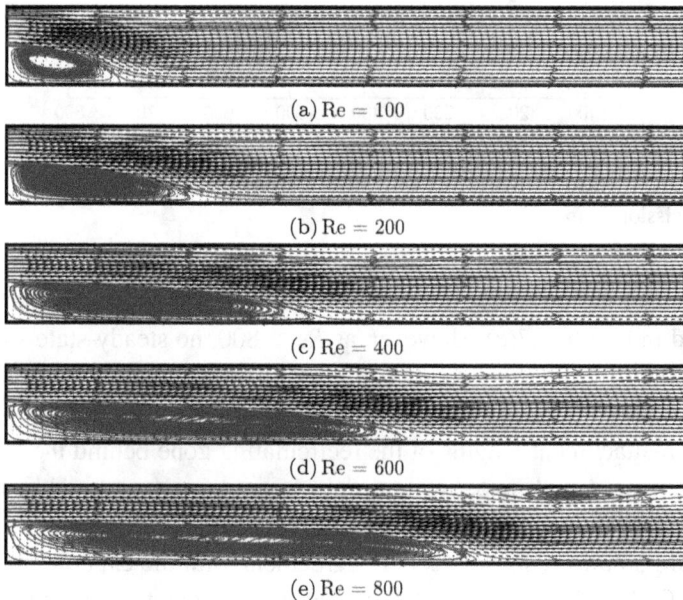

(a) Re $= 100$

(b) Re $= 200$

(c) Re $= 400$

(d) Re $= 600$

(e) Re $= 800$

Figure 5.7. Plots of streamlines for viscous Reynolds number (Reprinted with permission from [10]).

The boundary conditions adopted in this case are schematically illustrated in Figure 5.6.

One of the most significant observations made during this study is the formation of a vortex immediately behind the backstep. As the Reynolds number increases, the vortex's size increases proportionally, and this phenomenon is accurately captured by the incompressible GSM solver as shown in Figure 5.7. It is also found that when $Re = 800$, a secondary

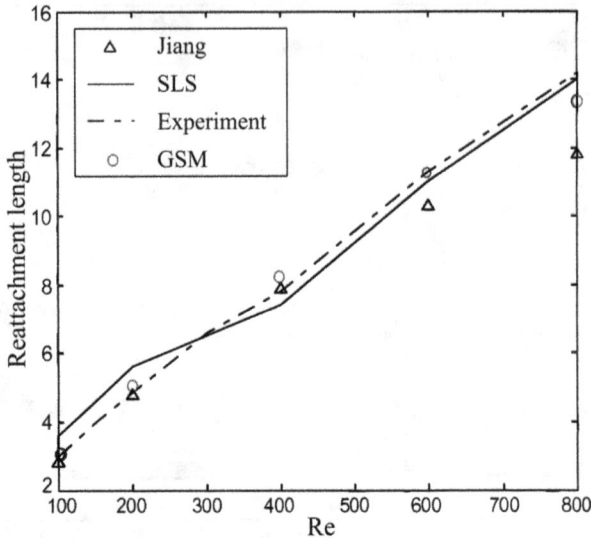

Figure 5.8. Predicted reattachment length ratios varied with Reynolds number (Reprinted with permission from [10]).

vortex attached to the top wall is induced in the downstream region, as depicted in Figure 5.7(e). However, as *Re* > 800, no steady-state solutions can be obtained in current practice, due mainly to the oscillation of induced vortices.

The reattachment lengths of the recirculating zone behind the backstep were compared with experimental data from Armaly *et al.* [11] and the least-squares finite element results from Jiang [12], as shown in Figure 5.8. The GSM solutions show excellent agreement with the experimental data, and the GSM solver provides the most accurate predictions among the different methods considered in this study.

5.3.3 *Steady-state lid-driven cavity flow*

The lid-driven cavity flow is another classical benchmark testing case designed to evaluate the behavior of algorithms that deal with incompressible viscous flows. In this test, incompressible viscous fluid is confined within a squared cavity, and only the upper edge is allowed to tangentially slide at a prescribed velocity ($u = 1.0$ in this study). Non-slip conditions

($u = v = 0$) are imposed on the rest of the walls, while the reference pressure is prescribed to be zero at the bottom left corner. The relevant boundary conditions used in the test are sketched in Figure 5.9(a). Two sets of grids, as shown in Figure 5.9(b) and 5.9(c), are studied: the coarser grid for Reynolds number flows ($Re \leq 3000$) and the finer grid for larger Reynolds number flows ($Re > 3000$), where the steeper boundary layers near walls must be resolved with finer grids.

The numerical results for the stream functions at various Reynolds numbers are shown in Figure 5.10. The transmission of momentum through viscosity produces a large-scale vortex in the center and small vortices at the corners. The vertical structure's details depend highly on the Reynolds number. For cases where $Re \leq 1$, the flow is almost symmetric with respect to the central vertical line, as shown in Figure 5.10(a). As the Reynolds number increases, the central vortex's position moves toward the top right corner before returning to the center of the cavity at higher Reynolds numbers. The bottom right and left vortices begin to develop at low Reynolds numbers, continuously increase in scale, and shift their position, as the Reynolds number increases. A secondary vortex is developed near the top left corner at approximately $Re = 3000$. Finally, at $Re = 10,000$ a new vortex is found to be developed at the right bottom corner. This case is considered as a limit for steady calculations, since a hop bifurcation appears for $Re > 10,000$. All the above-mentioned phenomena are well captured with the GSM solver, as shown in Figure 5.10.

Figure 5.11(a) and 5.11(b) shows, respectively, the profiles of u-component along the vertical line and v-component along a horizontal line through the center of the cavity with various Reynolds numbers. The incompressible GSM solutions show remarkably good agreements with referenced data from Ghia *et al.* [13].

5.3.4 *Time-dependent flow over a circular cylinder*

The GSM solver enhanced with the dual time-stepping approach and point-implicit RK45 method is applied to simulate the time-dependent flow past a circular cylinder. The test case follows the configuration as described in Section 5.3.1, but at $Re = 150$. The simulation is started with the following initial conditions: $u = 1.0, v = 0$, and $p = 0$.

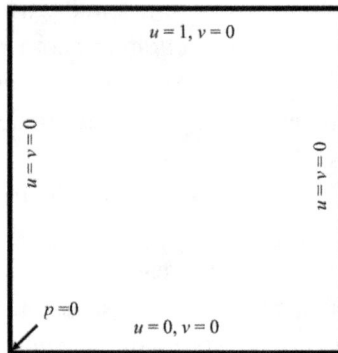

(a) boundary conditions

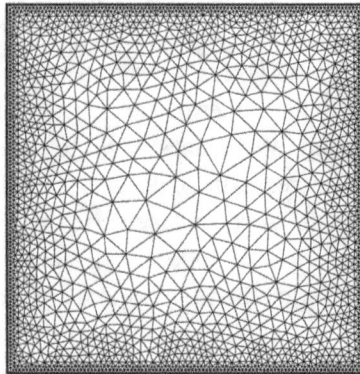

(b) coarse grid $(n_{node} = 2300)$

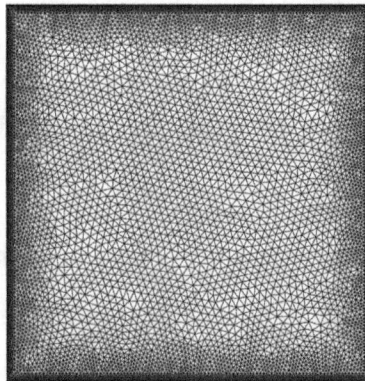

(c) fine grid $(n_{node} = 7771)$

Figure 5.9. Boundary conditions and grids studied in the lid-driven cavity flow problem (Reprinted with permission from [5]).

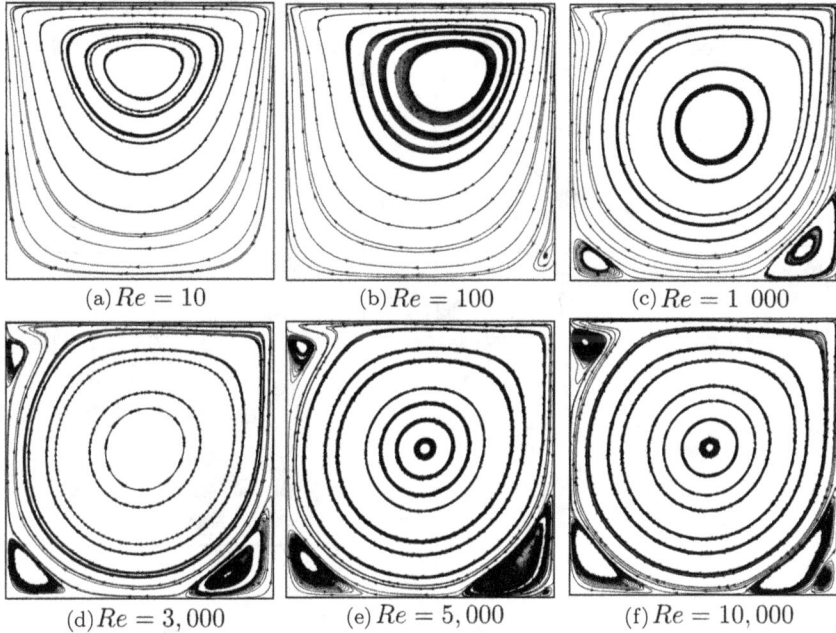

Figure 5.10. Plots of streamlines for various Reynolds number (Reprinted with permission from [10]).

A periodic flow pattern, known as the Von Karman vortex street, is finally captured with the GSM solver. Within a typical cycle of interest, a vortex originates from the upper rear portion of the cylinder surface and grows in scale before it is shed out of the surface. Subsequently, another vortex starts to generate on the lower rear portion of the cylinder surface, grows in scale, and is finally shed out of the surface and migrates downstream. Such a pair of vortices is periodically generated and released into the flow field, and eventually the well-known Von Karman vortex street is formed in the flow downstream of the cylinder.

The evolution of periodic flows in terms of streamlines and contours of pressure is shown in Figure 5.12, and the evolution of flows in terms of contours of velocity component u can be found in Figure 5.13, within a typical vortex-shedding period. Figure 5.14 shows the historical changes of drag and lift coefficients in some stable cycles. The oscillating behavior is fully related to the vortex-shedding change. Predicated lift and drag coefficients,

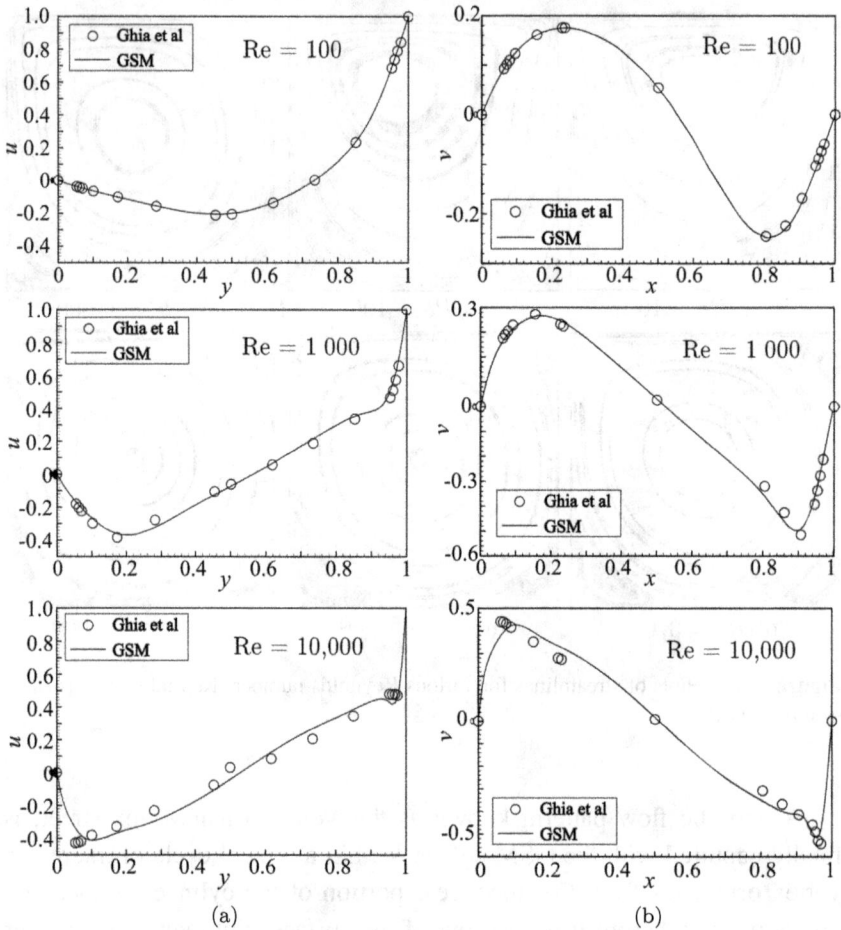

Figure 5.11. Profiles of *u*-component along the vertical line $x = 0.5$ (a) and *v*-component along the horizontal line $y = 0.5$ (b) for various Reynolds numbers (Reprinted with permission from [5]).

as well as the Strouhal number, are also summarized in Table 5.2. Some experimental data and relevant numerical results with different methodologies are also included for comparison. It is apparent that the GSM solver yields reasonably accurate results.

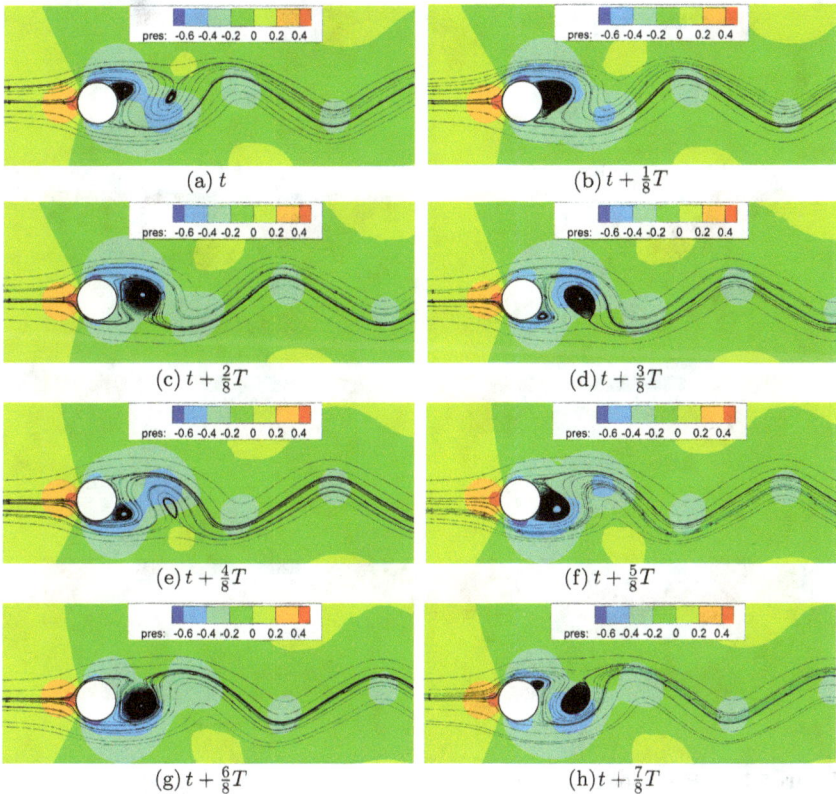

Figure 5.12. Evolution of flow field in terms of pressure contour and streamlines within a typical shedding cycle (Reprinted with permission from [10]).

The GSM has been successfully applied to solve incompressible flows. With the inclusion of artificial compressibility terms, the augmented N-S equations possess hyperbolic–parabolic properties such that either steady-state or unsteady solutions can be obtained with the help of the time marching approach. Besides, the dual time-stepping approach is adopted in GSM to couple with the point-implicit RK45 method for unsteady solutions. This study has demonstrated that the GSM is accurate, robust, and stable for incompressible flow problems.

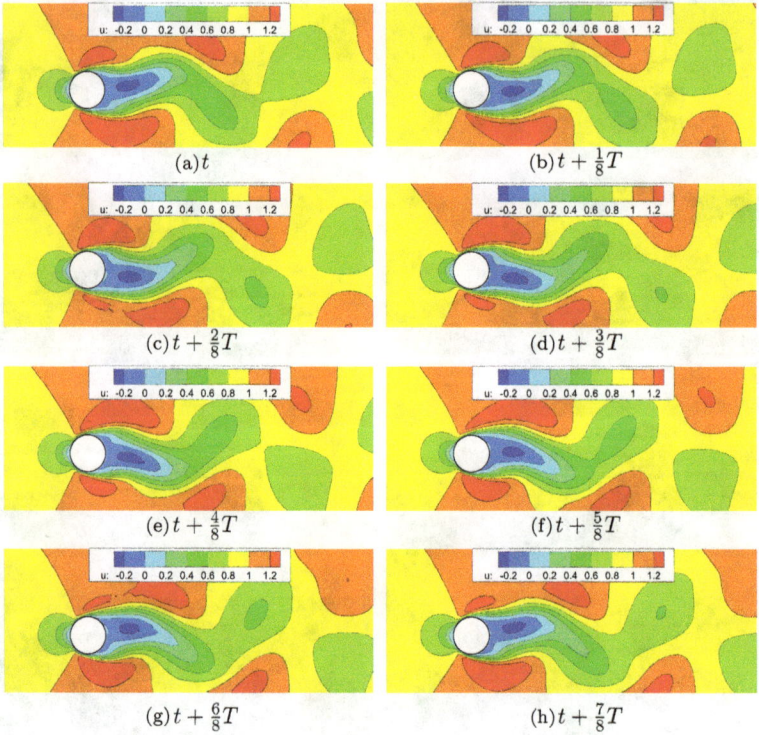

Figure 5.13. Evolution of flow field in terms of u-velocity within a typical shedding cycle (Reprinted with permission from [10]).

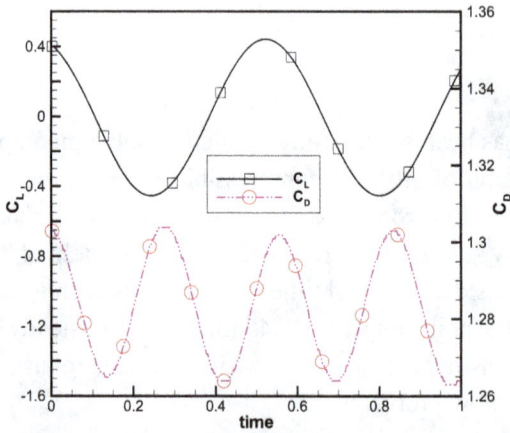

Figure 5.14. Variation of drag and lift coefficients (C_D and C_L) with time (Reprinted with permission from [5]).

Table 5.2. Comparison of predicted C_D, C_L, and St at $Re = 150$. The "Exp-1" is from [14], "Exp-2" is from [15], "Polar grid" is from [16]. "Triangular mesh" is from [17], and "Hybrid mesh" is from [18].

Reference	C_D	C_L	St
Exp-1	—	—	0.18
Exp-2	1.33	—	—
Polar grid	1.168 ± 0.025	± 0.486	0.18
Triangular mesh	1.166 ± 0.023	± 0.477	0.18
Hybrid mesh	1.380 ± 0.027	± 0.561	0.19
Triangular mesh (GSM)	1.283 ± 0.02	± 0.450	0.18

5.4 Concluding remarks

On comparing the GSM with the standard FDM, is the following points are noticed:

- The spatial derivatives in FDM are approximated using Taylor series expansion, while in the GSM, the gradient smoothing operator is utilized. After the derivatives are approximated, the follow-up treatments in GSM and FDM are essentially the same.
- The existing stability criteria developed for FDM are also applicable to GSM frameworks.
- Usually, standard FDM works well for structured meshes only. FDM can be made for less regular domains via domain transformation. However, there exist problems and limitations with such a transformation, and it is far from efficient for complicated problem domains. On the contrary, the GSM works very well for unstructured mesh with irregular triangular cells and hence is applicable to domains of arbitrarily complicated geometry. No transformation is needed for GSM, and naturally there are no related complications.
- When regular mesh is used and a proper set of GSDs is used, the GSM becomes an FDM. Therefore, FDM can be viewed as a special case of GSM in this regard. As the same mesh is employed in GSM as FDM, GSM holds the same 2nd-order accuracy as FDM.

The comparison of FDM and GSM is summarized in Table 5.3.

Table 5.3.　Comparison of FDM and GSM.

	FDM	GSM
Grid	Structured, no need for mesh generation	Unstructured, but needs mesh generation
Application to complex geometry	Challenging	Very natural
Accuracy	$\geq$ 2nd order for regular grid Inapplicable for irregular grid	$\geq$ 2nd order for regular grid $\geq$ 1st order for irregular grid
Stability	Good	Good
Efficiency	Faster	Fast
Implementation	Easier	Relatively hard

As a consequence, we can conclude that GSM is an innovative and unique numerical method with distinct features, and possesses strong potential in successful applications for versatile fluid flow problems.

References

[1] A. J. Chorin, A numerical method for solving incompressible viscous flow problems, *Journal of Computational Physics*, 135(2), 118–125 (1997, August). doi: 10.1006/jcph.1997.5716.

[2] P. L. Roe, Approximate Riemann solvers, parameter vectors, and difference schemes, *Journal of Computational Physics*, 43(2), 357–372 (1981, October 1). doi: 10.1016/0021-9991(81)90128-5.

[3] S. Shin, *Reynolds-Averaged Navier-Stokes Computation of Tip Clearance Flow in a Compressor Cascade Using an Unstructured Grid*. PhD thesis, Virginia Polytechnic Institute and State University, Virginia Tech (2001, September).

[4] D. L. Whitfield and L. Taylor, Numerical Solution of the Two-Dimensional Time-Dependent Incompressible Euler Equations. Technical Report NASA-CR-195775, (1994, April). Accessed (2020, February 27) Available: https://ntrs.nasa.gov/search .jsp?R=19940025355

[5] X. Xu, Development of Gradient Smoothing Method (GSM) for fluid flow problems, *National University of Singapore* (2009). doi: https://scholarbank.nus.edu.sg/handle/ 10635/160959.

[6] D. Kwak, J. L. C. Chang, S. P. Shanks, and S. R. Chakravarthy, A three-dimensional incompressible Navier-Stokes flow solver using primitive variables, *AIAA Journal*, 24(3), 390–396 (1986, May). doi: 10.2514/3.9279.

[7] S. Taneda, Experimental Investigation of the Wakes behind Cylinders and Plates at Low Reynolds Numbers, *Journal of the Physical Society of Japan*, 11(3), 302–307 (1956). doi: 10.1143/JPSJ.11.302.

[8]   Y. Huang, J. Deng, and A. Ren, Research on lift and drag in unsteady viscous flow around circular cylinders, *Journal of Zhejiang University*, 37, 596–601 (2003).

[9]   D. J. Tritton, Experiments on the flow past a circular cylinder at low Reynolds numbers, *Journal of Fluid Mechanics*, 6(4), 547–567 (1959). doi: 10.1017/S0022112059000829.

[10] G. X. Xu, E. Li, V. Tan, and G. R. Liu, Simulation of steady and unsteady incompressible flow using gradient smoothing method (GSM), *Computers & Structures*, 90–91, 131–144 (2012, January).

[11] B. F. Armaly, F. Durst, J. C. F. Pereira, and B. SchöNung, Experimental and theoretical investigation of backward-facing step flow, *Journal of Fluid Mechanics*, 127, 473–496, (1983). doi: 10.1017/S0022112083002839.

[12] B. -N Jiang, A least-squares finite element method for incompressible Navier-Stokes problems, *International Journal for Numerical Methods in Fluids*, 14(7), 843–859 (1992, April). doi: 10.1002/fld.1650140706.

[13] U. Ghia, K. N. Ghia, and C. T. Shin, High-Re Solutions for Incompressible Flow Using the Navier-Stokes Equations and a Multigrid Method*, *Journal of Computational Physics*, 48, 387–411 (1982).

[14] C. H. K. Williamson, Defining a universal and continuous Strouhal-Reynolds number relationship for the laminar vortex shedding of a circular cylinder, *Physics of Fluids*, 31(10), 2742–2744 (1988, October). doi: 10.1063/1.866978.

[15] E. Relf, Discussion of results of measurements of the resistance of wires, with some additional tests on the resistance of wires of small diameter. Technical Report Technical Report 102. The advisory committee for aeronautics (1915).

[16] A. Belov, L. Martinelli, and A. Jameson, a new implicit algorithm with multigrid for unsteady incompressible flow calculations. *Aerospace Research Central* (2012). doi: 10.2514/6.1995-49.

[17] P. Lin, L. Martinelli, T. J. Baker, and A. Jameson, Two-dimensional implicit time dependent calculations for incompressible flows on adaptive unstructured meshes, in *15th AIAA Computational Fluid Dynamics Conference* (2001). doi: 10.2514/6.2001-2655.

[18] Y. Kallinderis and H. T. Ahn, "Incompressible Navier-Stokes method with general hybrid meshes," *Journal of Computational Physics*, 210(1), 75–108 (2005, November). doi: 10.1016/j.jcp.2005.04.002.

Chapter 6

Theory and Formulation for Lagrangian GSM

Contents

This chapter presents the theory of Lagrangian GSM (L-GSM) for solving large deformation problems with moving particles. The key formulations and techniques are discussed in detail. The accuracy performance and stability of the L-GSM are investigated through a number of numerical tests.

This chapter is outlined as follows:

- Section 6.1 sets important assumptions for L-GSM.
- The key techniques for L-GSM, including the gradient smoothing operator, construction of GSD, free surface treatment, particle connecting mechanism, numerical oscillation suppression, conservation issues, boundary treatments, and time integration scheme, are presented in Sections 6.2–6.9.
- Sections 6.10 and 6.11 assess the performance of L-GSM in terms of accuracy and stability, respectively, via numerical tests on benchmark problems.

Why L-GSM?

SPH is a Lagrangian meshfree method that can solve strong-form PDEs governing fluids and solids undergoing extremely large deformation [1–4]. It has advantages over the standard FEM [5] for solving solid mechanics problems. It also suffers from two inherent issues: tensile instability rooted in the gradient approximation used in SPH [6–8] and low computation efficiency due to the necessary update of supporting partices at every time step. Both issues are rooted in the gradient approximation method used in SPH.

On the other hand, GSM has shown outstanding performance in accuracy and stability, as demonstrated in the previous chapters. L-GSM [9, 10] is a combination of SPH and GSM that aims to resolve these issues using the gradient smoothing technique in a more controllable manner, leading to better accuracy and stability while also improving efficiency. This is because the rigorously formulated gradient smoothing operations needs much fewer supporting particles for accurate and stable approximation. Moreover, L-GSM works with the Lagrangian frames and uses automatically generatable triangular grids/mesh, reducing the need for frequent particle connection updates, which also contributes to better efficiency.

6.1 Assumptions used in L-GSM

Assumption 1: Particle representation. The L-GSM method assumes that the problem domain is divided into a finite set of particles, each with field variables such as mass, density, stress, strain, location, velocity, and energy.

Assumption 2: Lagrangian particle. These particles are considered Lagrangian, meaning they have a constant mass m and do not transfer mass between each other. This ensures that mass conservation is preserved throughout the simulation.

Assumption 3: Volume decoupling. L-GSM uses a technique called volume decoupling, where each particle has two sets of volume and density values. The first is the real control volume (V^{real}) with corresponding density (ρ^{real}) from the continuity equation with a relationship of $V^{real}\rho^{real} = m$. The second is the nominal volume ($V^{nominal}$) used for numerical operations, with corresponding density ($\rho^{nominal}$) that satisfies $V^{nominal}\rho^{nominal} = m$. The nominal volume is the local gradient smoothing domain constructed for gradient smoothing, as explained in Chapter 2. While these two volumes may not be exactly equal, they are typically very close. This is because, similar to SPH, L-GSM does not apply constraints onto the relation between particles' motion and density. This volume-decoupling technique enables L-GSM to have both features of SPH and GSM.

Assumption 4: Governing equations. The L-GSM method governs the particles' interaction and motion in Lagrangian frame through the

Navier–Strokes PDEs, same as in the SPH method.

$$\begin{cases} \dfrac{D\rho}{Dt} = -\rho \dfrac{\partial v^\beta}{\partial x^\beta} \\[2mm] \dfrac{Dv^\alpha}{Dt} = \dfrac{1}{\rho} \dfrac{\partial \sigma^{\alpha\beta}}{\partial x^\beta} + F^\alpha_{external} \\[2mm] \dfrac{Dd^\alpha}{Dt} = v^\alpha \end{cases} \tag{6.1}$$

These equations use superscript Greek indices α and β, which denote the Cartesian components x, y, z, and the Einstein convection is applied to the repeated Greek indices. The variables include density (ρ), velocity (v), displacement (d), total stress tensor ($\sigma^{\alpha\beta}$), and acceleration may be caused by external forces ($F_{external}$) like gravity.

The GSM approximates the spatial differential terms in the equations. The continuity equation yields the real density ρ^{real}, while the nominal density is obtained directly from $\rho^{nominal} = \frac{m}{V^{nominal}}$. Using nominal density in computation can conserve momentum precisely, which will be explained in Section 6.4. For simplicity purposes, the energy equation is not included, but it can be added if necessary. The temporal differential terms are dealt with in the same way as in the SPH, using a standard explicit time integration scheme.

Assumption 5: *Gradient approximation.* The smoothed gradients, as computed in GSM, are plugged directly into the governing PDEs as done in FDM. This provides L-GSM with the advantages of simplicity in implementation. Domain integration, as in the standard FEM [5], is not required.

6.2 Gradient smoothing for L-GSM

6.2.1 *For 2D problems*

To compute the smoothed gradient of an arbitrary field variable, say U, at particle i in L-GSM, various gradient smoothing domains (GSDs) can be used, as discussed in Chapter 2. The simplest form, known as the node-based GSD or nGSD, is used in this chapter. An nGSD is constructed locally around particle i, as shown in Figure 6.1. The two components of

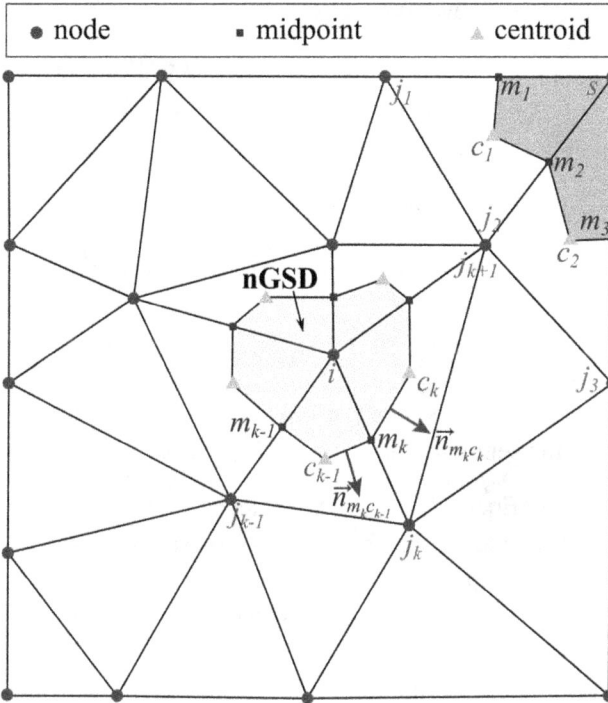

Figure 6.1. A polygonal nGSD with 10 segments for computing the smoothed gradient for inner particle i, and an nGSD for boundary particle s.

the smoothed gradient are then computed using the following equation:

$$\begin{cases} \dfrac{\partial U_i}{\partial x} = \dfrac{1}{V_i} \displaystyle\sum_{k=1}^{n_i} \dfrac{U_i + U_{j_k}}{2} (\Delta S_x)_{ij_k}, \\[4mm] \dfrac{\partial U_i}{\partial y} = \dfrac{1}{V_i} \displaystyle\sum_{k=1}^{n_i} \dfrac{U_i + U_{j_k}}{2} (\Delta S_y)_{ij_k} \end{cases} \tag{6.2}$$

where

- $(U_i + U_{j_k})/2$ is the average value on the two boundary segments $c_{k-1}m_k$ and $m_k c_k$ related to the supporting particle j_k, based on the rectangular rule.
- $(\Delta S_x)_{ij_k}$ and $(\Delta S_y)_{ij_k}$ represent the rate of the change of U resulting from the geometry change of the nGSD segment related to edge-connecting

particles i and j_k. They have the form of

$$\begin{cases} (\Delta S_x)_{ij_k} = (\Delta S_x)_{ij_k}^{(L)} + (\Delta S_x)_{ij_k}^{(R)} \\ (\Delta S_y)_{ij_k} = (\Delta S_y)_{ij_k}^{(L)} + (\Delta S_y)_{ij_k}^{(R)} \end{cases} \tag{6.3}$$

$$\begin{cases} (\Delta S_x)_{ij_k}^{(L)} = r_{m_k c_k}(n_x)_{m_k c_k}, \quad (\Delta S_y)_{ij_k}^{(L)} = r_{m_k c_k}(n_y)_{m_k c_k} \\ (\Delta S_x)_{ij_k}^{(R)} = r_{m_k c_{k-1}}(n_x)_{m_k c_{k-1}}, \quad (\Delta S_y)_{ij_k}^{(R)} = r_{m_k c_{k-1}}(n_y)_{m_k c_{k-1}} \end{cases} \tag{6.4}$$

where

- r is the length of the segment;
- n_x and n_y are the components of outward unit vector **n** (and hence the direction of the segment), as shown in Figure 6.1.

The summation in Eq. (6.2) counts the collective segmental changes for the whole nGSD constructed for particle i. All these formulas are for spatial discretization and are the same as those detailed in Chapter 2.

The term $r_{m_k c_k}(n_x)_{m_k c_k}$ in Eq. (6.4) is the component of vector $\overrightarrow{m_k c_k}$ in y direction. Using Eqs. (2.19)–(2.22), we have

$$r_{m_k c_k}(n_x)_{m_k c_k} = y_{c_k} - y_{m_k} \tag{6.5}$$

$$r_{m_k c_k}(n_y)_{m_k c_k} = x_{c_k} - x_{m_k} \tag{6.6}$$

$$r_{c_{k-1} m_k}(n_x)_{c_{k-1} m_k} = y_{m_k} - y_{c_k} \tag{6.7}$$

$$r_{c_{k-1} m_k}(n_y)_{c_{k-1} m_k} = x_{m_k} - x_{c_k} \tag{6.8}$$

By simple geometry, we have

$$x_{c_k} = \frac{x_i + x_{j_k} + x_{j_{k+1}}}{3}, \quad y_{c_k} = \frac{y_i + y_{j_k} + y_{j_{k+1}}}{3} \tag{6.9}$$

$$x_{m_k} = \frac{x_i + x_{j_k}}{2}, \quad y_{m_k} = \frac{y_i + y_{j_k}}{2} \tag{6.10}$$

Substituting Eqs. (6.4)–(6.10) into (6.2) yields

$$\begin{cases} \dfrac{\partial U_i}{\partial x} = \dfrac{1}{6V_i} \sum_{k=1}^{n_i} (U_i + U_{j_k})(y_{j_{k+1}} - y_{j_{k-1}}), \\ \dfrac{\partial U_i}{\partial y} = \dfrac{1}{6V_i} \sum_{k=1}^{n_i} (U_i + U_{j_k})(x_{j_{k+1}} - x_{j_{k-1}}). \end{cases} \tag{6.11}$$

We see now, in L-GSM, the gradient of a field variable U at a given particle i is expressed in terms of the field variable and the coordinates of the particles that surround particle i. This is similar to the SPH equation for particle approximation. However, unlike in SPH, the particles used in L-GSM are fixed for a given particle, meaning that the number of particles involved is much less, and particle searching need not be done frequently. For instance, in Figure 6.1, the particle i is surrounded by a total of six particles, including itself, which is roughly half or even one-third of the particles used in SPH.

6.2.2 For 3D problems

The equation for computing the smoothed gradient for particle i in a 3D domain using the nGSD is given as

$$
\begin{cases}
\dfrac{\partial U_i}{\partial x} = \dfrac{1}{V_i} \sum_{k=1}^{n_i} \dfrac{U_i + U_{j_k}}{2} (\Delta S_x)_{ij_k}, \\[2ex]
\dfrac{\partial U_i}{\partial y} = \dfrac{1}{V_i} \sum_{k=1}^{n_i} \dfrac{U_i + U_{j_k}}{2} (\Delta S_y)_{ij_k}, \\[2ex]
\dfrac{\partial U_i}{\partial z} = \dfrac{1}{V_i} \sum_{k=1}^{n_i} \dfrac{U_i + U_{j_k}}{2} (\Delta S_z)_{ij_k}.
\end{cases}
\tag{6.12}
$$

In this case, the smoothed gradient for a field variable U at a target particle i is computed by evaluating U on the boundary of the nGSD of particle i. The nGSD surface is composed of quadrilateral surface segments, and $(\Delta S_x)_{ij_k}, (\Delta S_y)_{ij_k}$, and $(\Delta S_z)_{ij_k}$ in 3D case are given by the following equations:

$$
\begin{cases}
(\Delta S_x)_{ij_k} = \sum_{l=1}^{L} A_{ij_k l} (n_x)_{ij_k l}, \\[2ex]
(\Delta S_y)_{ij_k} = \sum_{l=1}^{L} A_{ij_k l} (n_y)_{ij_k l}, \\[2ex]
(\Delta S_z)_{ij_k} = \sum_{l=1}^{L} A_{ij_k l} (n_z)_{ij_k l}
\end{cases}
\tag{6.13}
$$

where $A_{ij_k l}$ is the area of the quadrilateral, L denotes the total number of quadrilaterals associated with the supporting particle j_k, while n_x, n_y, and

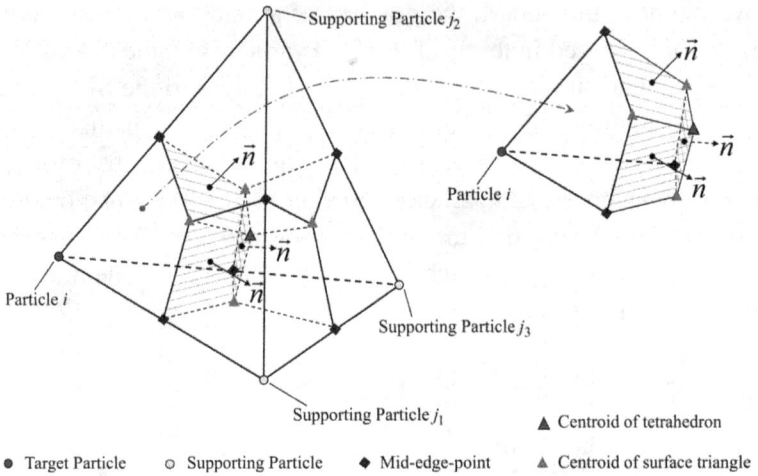

Figure 6.2. A part of nGSD for particle i located within one tetrahedral cell for computing the smoothed gradient. The surface of this part of nGSD consists of three quadrilateral surfaces, each of which has an outward unit normal.

n_z are the components of outward unit vector $\mathbf{n}$ of the quadrilateral surface in the x, y, and z directions, respectively, as dipicted in Fig. 6.2.

Using Eq. (6.2) in 2D and Eq. (6.12) in 3D, the smoothed gradients of velocity and stress can be obtained, and the general form of the N-S equations is discretized as

$$\begin{cases} \dfrac{D\rho}{Dt} = -\dfrac{\rho_i}{V_i} \sum_{k=1}^{n_i} \left(\dfrac{v_{j_k}^\alpha + v_i^\alpha}{2} \right) (\Delta S_\alpha)_{ij_k} \\[2ex] \dfrac{Dv^\alpha}{Dt} = \dfrac{1}{\rho_i V_i} \sum_{k=1}^{n_k} \left(\dfrac{\sigma_{j_k}^{\alpha\beta} + \sigma_i^{\alpha\beta}}{2} \right) (\Delta S_\alpha)_{ij_k} + F^\alpha \\[2ex] \dfrac{Dd^\alpha}{Dt} = v^\alpha \end{cases} \qquad (6.14)$$

It is seen that this set of discretized system equations is very similar to that of SPH.

6.2.3 Accuracy analysis of the smoothed gradient

Consider 2D cases: To evaluate the accuracy of the smoothed gradient in Eq. (6.2), we can use Taylor's expansion to approximate the field variable

U at point j_k, in reference to particle i.

$$U_{j_k} = U_i + (x_{j_k} - x_i)\frac{\partial U_i}{\partial x} + (y_{j_k} - y_i)\frac{\partial U_i}{\partial y} + O(\Delta r^2) \tag{6.15}$$

where the error term is at the 2nd order of the spatial step length Δr, and takes the form of a quadratic equation in terms of $(\Delta x)^2, (\Delta y)^2, (\Delta x \Delta y)$. Take $\partial U / \partial x$ as an example. Substitution of Eq. (6.15) into Eq. (6.11) gives

$$\frac{\partial U_i}{\partial x}\bigg|_{GSM} = A_1 U_i + B_1 \frac{\partial U_i}{\partial x} + C_1 \frac{\partial U_i}{\partial y} + O(\Delta r) \tag{6.16}$$

where A_1, B_1, and C_1 are given by

$$\begin{cases} A_1 = \dfrac{1}{V_i} \displaystyle\sum_{k=1}^{n_i} \dfrac{(y_{j_{k+1}} - y_{j_{k-1}})}{3}; \\[3mm] B_1 = \dfrac{1}{V_i} \displaystyle\sum_{k=1}^{n_i} \dfrac{(y_{j_{k+1}} - y_{j_{k-1}})(x_{j_k} - x_i)}{6}; \\[3mm] C_1 = \dfrac{1}{V_i} \displaystyle\sum_{k=1}^{n_i} \dfrac{(y_{j_{k+1}} - y_{j_{k-1}})(y_{j_k} - y_i)}{6}. \end{cases} \tag{6.17}$$

Note the reduction in the order of accuracy in Eq. (6.16). This is because V_i is at the 2nd order of the step length Δr, and the term $(\Delta S_x)_{ij_k}$ is at the first order of Δr. As a result, the error term becomes first order of Δr, causing a reduction in accuracy.

$$\frac{(\Delta S_x)_{ij_k}}{V_i} O(\Delta r^2) \approx O(\Delta r) \tag{6.18}$$

Thus, when approximating the first derivative, the accuracy naturally decreases by one order compared to approximating the function itself.

To ensure that our GSM accurately produces $\partial U / \partial x$, the coefficients must satisfy $A_1 = 0$, $B_1 = 1$, and $C_1 = 0$. Once these conditions are met, the resulting smoothed gradient will be at least 1st order accurate.

Consider an inner particle i in Figure 6.1. It has

$$\sum_{k=1}^{n_i} y_{j_{k+1}} \equiv \sum_{k=1}^{n_i} y_{j_k} \equiv \sum_{k=1}^{n_i} y_{j_{k-1}} \tag{6.19}$$

We found

$$A_1 = \frac{1}{3V_i}\left(\sum_{k=1}^{n_i} y_{j_{k+1}} - \sum_{k=1}^{n_i} y_{j_{k-1}}\right) = 0 \tag{6.20}$$

That is, the requirement $A_1 = 0$ is satisfied.

In examining B_1, we have

$$B_1 = \frac{1}{6V_i}\left(\sum_{k=1}^{n_i} y_{j_{k+1}} x_{j_k} - \sum_{k=1}^{n_i} x_{j_k} y_{j_{k-1}}\right)$$

$$= \frac{1}{6V_i}\left(\sum_{k=1}^{n_i} y_{j_{k+1}} x_{j_k} - \sum_{k=1}^{n_i} x_{j_{k+1}} y_{j_k}\right) \tag{6.21}$$

In matrix form,

$$B_1 = \frac{1}{3V_i}\left(\frac{1}{2}\sum_{k=1}^{n_i}\begin{vmatrix} x_{j_k} & y_{j_k} & 1 \\ x_{j_{k+1}} & y_{j_{k+1}} & 1 \\ 0 & 0 & 1 \end{vmatrix}\right) = \frac{\sum_{k=1}^{n_i} S_{\Delta O j_k j_{k+1}}}{3V_i} \tag{6.22}$$

where

- $S_{\Delta O j_k j_{k+1}}$ is the area of the triangle made of points origin O, point j_k, and point j_{k+1}.
- The sum of areas $\sum_{k=1}^{n_i} S_{\Delta O j_k j_{k+1}}$ is the total area of the polygon formed by the supporting particles of particle i, which is triple of the area V_i of the gradient smoothing domain of particle i.

Hence, $B_1 = 1$ holds.

Similar to B_1, we find,

$$C_1 = \frac{1}{6V_i}\left(\sum_{k=1}^{n_i} y_{j_{k+1}} y_{j_k} - \sum_{k=1}^{n_i} y_{j_{k-1}} y_{j_k}\right)$$

$$= \frac{1}{6V_i}\left(\sum_{k=1}^{n_i} y_{j_k} y_{j_{k-1}} - \sum_{k=1}^{n_i} y_{j_{k-1}} y_{j_k}\right) = 0 \tag{6.23}$$

After satisfying the conditions $A_1 = 0$, $B_1 = 1$, and $C_1 = 0$, we can state that the smoothed gradient given in Eq. (6.2) is consistent or at least 1st order accurate in all cases, for the inner particle i as shown in Figure 6.1.

This consistency is crucial because it enables the seamless integration of nGSD with the GSM gradient smoothing. By following the same derivations presented in Eqs. (6.15)–(6.23), we can also obtain the same conclusion for the 3D smoothed gradient given in Eq. (6.12).

The key difference between L-GSM and SPH lies in the consistency of gradient calculations with linear field variable approximations. L-GSM ensures stability, convergence, and conservation, even in unstructured grids, through an orthogonal projection. It is *error free* in the gradient calculation of constant and linear functions. For constant functions, the field function U_i in Eq. (6.2) gives the same form as the term A_1 in Eq. (6.17), which has been proven to be zero. For linear functions such as x, the smoothed gradient is equivalent to the term B_1 in Eq. (6.17), which has been shown to be equal to 1. In contrast, SPH can only achieve 0th-order consistency in its particle approximation form and requires an additional proper normalization.

The accuracy condition of 2D and 3D smoothed gradients will be investigated numerically and compared to the corresponding SPH gradient operator in Section 6.10.2.

6.3 Free surface treatments for L-GSM

For an L-GSM model, there are two types of particles: particles inside the problem domain (referred to as "inner particles", i) and particles on the boundary (referred to as "free surface particles", s), as shown in Figure 6.1. The smoothed gradient equations (Eqs. (6.2) and (6.12)) are derived for inner particles only based on the relationship in Eq. (6.19). Thus, it cannot produce an accurate gradient for free surface particles, which can be seen through the following analysis.

For free surface particles, we have $(\Delta S_x)_{ij_1}^L = 0$, $(\Delta S_y)_{ij_1}^L = 0$ and $(\Delta S_x)_{ij_3}^R = 0$, $(\Delta S_y)_{ij_3}^R = 0$, which leads to

$$A_1 = \frac{1}{V_i} \sum_{k=1}^{n_i} (\Delta S_x)_{ij_k} = \frac{1}{V_i}(y_{m_3} - y_{m_1}) \neq 0 \qquad (6.24)$$

Similarly, it is easy to find that $B_1 \neq 1$ and $C_1 \neq 0$ for free surface particles. This suggests that the smoothed gradient cannot produce an accurate gradient for free surface particles.

To restore the consistency for these free surface particles, two approaches have been proposed: the consistent form of gradient smoothing on free surface and the normalized form of gradient smoothing [9].

6.3.1 *Consistent form of gradient smoothing on free surface*

This treatment makes use of the free surface condition, i.e., the stress on the free surface should be zero pressure or zero tractions while the velocity shall equal that of the related free surface particle, as illustrated in Figure 6.3. The term "consistent" in this context refers to a form of gradient smoothing that is consistent with the ideal free surface conditions.

For example, consider a free surface particle labeled "s" in Figure 6.3, which is bounded by 6 nGSD boundary segments, including two free surface segments. On these free surface segments, the pressure value is set to be zero, and the velocity value is set equal to the velocity of the free surface particle "v_s". By incorporating these boundary conditions into the gradient smoothing equations, it is possible to achieve a

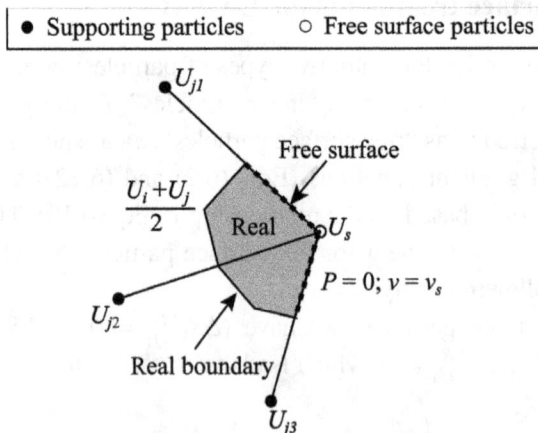

Figure 6.3. Sketch of supporting domain for free surface particle s.

more accurate and consistent representation of the free surface particle's behavior.

The free surface boundary treatment is employed in the application to problems with a free surface.

6.3.2 *Normalized form of gradient smoothing*

For more general mathematical problems that involve boundaries, the gradient smoothing on boundary particles can be approximated by normalizing the smoothed gradient.

For 2D cases, applying the Taylor's expansion (6.15) to the term $\frac{1}{V_i}\sum_{i=1}^{n_i}(y_{j_{k+1}}-y_{j_{k-1}})(U_{j_k}-U_i)$ yields

$$
\begin{aligned}
\frac{1}{V_i}\sum_{k=1}^{n_i} & \frac{(y_{j_{k+1}}-y_{j_{k-1}})(U_{j_k}-U_i)}{6} \\
&= \left[\frac{1}{V_i}\sum_{k=1}^{n_i}\frac{(y_{j_{k+1}}-y_{j_{k-1}})(x_{j_k}-x_i)}{6}\right]\frac{\partial U_i}{\partial x} \\
&\quad + \left[\frac{1}{V_i}\sum_{k=1}^{n_i}\frac{(y_{j_{k+1}}-y_{j_{k-1}})(y_{j_k}-y_i)}{6}\right]\frac{\partial U_i}{\partial y}+O(\Delta r)
\end{aligned}
\tag{6.25}
$$

Note the loss of accuracy for the same reason given in Eq. (6.18). Equation (6.25) can be written as

$$
A = \Sigma_1 U_i^x + \Sigma_2 U_i^y + O(\Delta r)
\tag{6.26}
$$

with

$$
A = \frac{1}{V_i}\sum_{k=1}^{n_i}\frac{(y_{j_{k+1}}-y_{j_{k-1}})(U_{j_k}-U_i)}{6}
\tag{6.27}
$$

$$
\Sigma_1 = \frac{1}{V_i}\sum_{k=1}^{n_i}\frac{(y_{j_{k+1}}-y_{j_{k-1}})(x_{j_k}-x_i)}{6}
\tag{6.28}
$$

$$
\Sigma_2 = \frac{1}{V_i}\sum_{k=1}^{n_i}\frac{(y_{j_{k+1}}-y_{j_{k-1}})(y_{j_k}-y_i)}{6}
\tag{6.29}
$$

Similarly, applying the Taylor's expansion (6.15) to the term $\frac{1}{V_i}\sum_{i=1}^{n_i}$ $(x_{j_{k+1}} - x_{j_{k-1}})(U_{j_k} - U_i)$ yields

$$B = \Sigma_3 U_i^x + \Sigma_4 U_i^y + O(\Delta r) \tag{6.30}$$

with

$$B = \frac{1}{V_i}\sum_{k=1}^{n_i}\frac{(x_{j_{k+1}} - x_{j_{k-1}})(U_{j_k} - U_i)}{6} \tag{6.31}$$

$$\Sigma_3 = \frac{1}{V_i}\sum_{k=1}^{n_i}\frac{(x_{j_{k+1}} - x_{j_{k-1}})(x_{j_k} - x_i)}{6} \tag{6.32}$$

$$\Sigma_4 = \frac{1}{V_i}\sum_{k=1}^{n_i}\frac{(x_{j_{k+1}} - x_{j_{k-1}})(y_{j_k} - y_i)}{6} \tag{6.33}$$

Rewriting Eqs. (6.26) and (6.30) in matrix format yields

$$\begin{bmatrix} A \\ B \end{bmatrix} = \begin{bmatrix} \Sigma_1 & \Sigma_2 \\ \Sigma_3 & \Sigma_4 \end{bmatrix}\begin{bmatrix} U_i^x \\ U_i^y \end{bmatrix} + O(\Delta r) \tag{6.34}$$

Hence, we get

$$\begin{bmatrix} U_i^x \\ U_i^y \end{bmatrix} = \begin{bmatrix} \Sigma_1 & \Sigma_2 \\ \Sigma_3 & \Sigma_4 \end{bmatrix}^{-1}\begin{bmatrix} A \\ B \end{bmatrix} \tag{6.35}$$

Equation (6.35) is the "normalized" form of the smoothed gradient in 2D which restores at least 1st-order accuracy for all particles, including boundary particles on free surface. The coefficient matrix reflects the effective area or volume of the nGSD, and its inverse exists, as long as all these supporting particles (including s) are not on a line forming a non-zero volume.

The above steps (6.25)–(6.35) can be easily extended to the 3D case (see [11]) for detail:

$$\begin{bmatrix} U_i^x \\ U_i^y \\ U_i^z \end{bmatrix} = \begin{bmatrix} \Sigma_{11} & \Sigma_{12} & \Sigma_{13} \\ \Sigma_{21} & \Sigma_{22} & \Sigma_{23} \\ \Sigma_{31} & \Sigma_{32} & \Sigma_{33} \end{bmatrix}^{-1}\begin{bmatrix} A \\ B \\ C \end{bmatrix} \tag{6.36}$$

where

$$A = \frac{1}{V_i} \sum_{k=1}^{n_i} \frac{(U_{j_k} - U_i)(\Delta S_x)_{ij_k}}{2}$$

$$\Sigma_{11} = \frac{1}{V_i} \sum_{k=1}^{n_i} \frac{(x_{j_k} - x_i)(\Delta S_x)_{ij_k}}{2}$$

$$\Sigma_{12} = \frac{1}{V_i} \sum_{k=1}^{n_i} \frac{(y_{j_k} - y_i)(\Delta S_x)_{ij_k}}{2} \qquad (6.37)$$

$$\Sigma_{13} = \frac{1}{V_i} \sum_{k=1}^{n_i} \frac{(z_{j_k} - z_i)(\Delta S_x)_{ij_k}}{2}$$

$$B = \frac{1}{V_i} \sum_{k=1}^{n_i} \frac{(U_{j_k} - U_i)(\Delta S_y)_{ij_k}}{2}$$

$$\Sigma_{21} = \frac{1}{V_i} \sum_{k=1}^{n_i} \frac{(x_{j_k} - x_i)(\Delta S_y)_{ij_k}}{2}$$

$$\Sigma_{22} = \frac{1}{V_i} \sum_{k=1}^{n_i} \frac{(y_{j_k} - y_i)(\Delta S_y)_{ij_k}}{2} \qquad (6.38)$$

$$\Sigma_{23} = \frac{1}{V_i} \sum_{k=1}^{n_i} \frac{(z_{j_k} - z_i)(\Delta S_y)_{ij_k}}{2}$$

$$C = \frac{1}{V_i} \sum_{k=1}^{n_i} \frac{(U_{j_k} - U_i)(\Delta S_z)_{ij_k}}{2}$$

$$\Sigma_{31} = \frac{1}{V_i} \sum_{k=1}^{n_i} \frac{(x_{j_k} - x_i)(\Delta S_z)_{ij_k}}{2}$$

$$\Sigma_{32} = \frac{1}{V_i} \sum_{k=1}^{n_i} \frac{(y_{j_k} - y_i)(\Delta S_z)_{ij_k}}{2} \qquad (6.39)$$

$$\Sigma_{33} = \frac{1}{V_i} \sum_{k=1}^{n_i} \frac{(z_{j_k} - z_i)(\Delta S_z)_{ij_k}}{2}$$

For programming simplicity, the area V_i of nGSD in the expressions of A, B, C and $\Sigma's$ can be removed because it will be cancelled when calculating the inverse matrix in Eqs. (6.35) and (6.36).

Numerical experiments demonstrate that this treatment can also be used to mimic the free surface boundary conditions.

6.4 Supporting particle selections

In numerical methods, approximating field variables is one of the most critical operations. In grid-based methods such as FEM, a global mesh with nodes is used, and the function and its gradients are approximated based on elements. On the other hand, particle or meshfree methods like SPH generate a set of particles to represent the media, and the approximation is done using a set of locally selected supporting particles.

In L-GSM, supporting particles are also used, similar to in the SPH, but with an additional procedure of forming smoothing domains for computing the smoothed gradients. This can be done in two ways, as shown in Figure 6.4 (Mao *et al.*, 2019):

- *Global L-GSM*: It uses a *global* unstructured mesh with particles at the nodes, and the approximation and gradient smoothing are performed for each particle using only one layer of particles, as shown in Figure 6.4(a), where these 6 gray supporting particles are selected. This method uses a fixed set of supporting particles until the re-meshing is performed. This global L-GSM is the one we discussed so far.

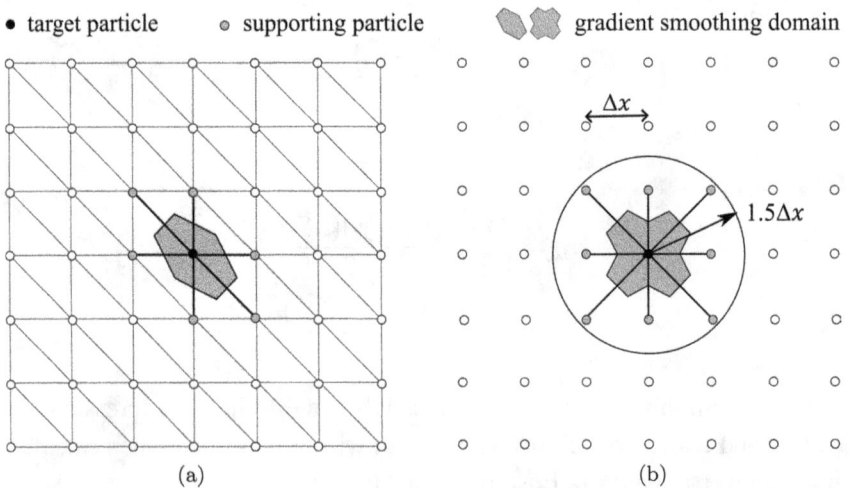

Figure 6.4. Smoothing domains formed using supporting particles. (a) Global L-GSM, (b) local L-GSM (Reprinted with permission from [10]).

- *Local L-GSM*: It uses a *local* unstructured mesh. It is created on the fly based on the currently given particle distribution, which is very similar to SPH, as shown in Figure 6.4(b). The supporting particles for the local L-GSM do not need to be updated frequently since the relative movement of particles is small under the rigid control of the critical time step.

Our study has shown that the use of either global or local mesh does not significantly affect the accuracy of gradient approximation. However, the global L-GSM can ensure conservation, while the local L-GSM can only approximate it.

6.4.1 *Global mesh and re-mesh*

In L-GSM, only an unstructured mesh is required. A mesh can be easily obtained using the well-established Delaunay triangulation algorithm [13, 14] or the advancing front techniques. However, in extreme events where the media undergo significant deformation, the mesh may become deformed with skinny triangles, as shown in Figure 6.5(a). Although the L-GSM may still work mathematically, the approximation may become unphysical since the behavior of a particle depends on particles far away.

To address this issue, the following modification is made on the Delaunay mesh. When the minimum internal angle of a triangle becomes smaller

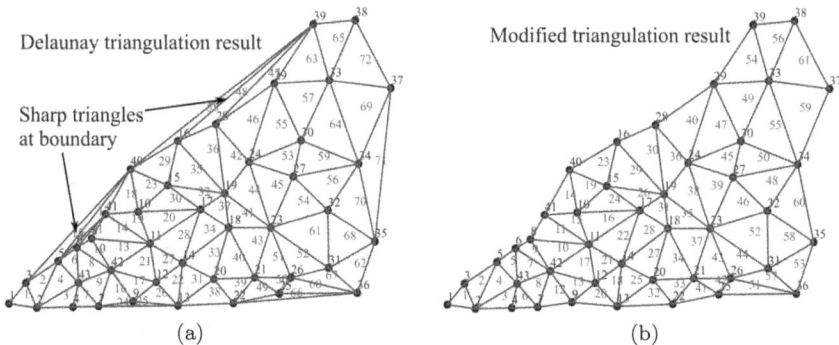

Figure 6.5. Examples of Delaunay mesh (a) and modified mesh (b).

than $\theta_{min} = 10°$ or the longest edge is longer than $L_{max} = 2.5\Delta r$, the triangle is removed, resulting in a cleaned-up mesh, as shown in Figure 6.5 (b). This approach helps to ensure that the mesh remains suitable for accurate and physically meaningful approximations, even in extreme deformation events.

6.4.2 *Local supporting particle selection algorithm*

The local L-GSM searches supporting particles with a simple and efficient algorithm called local supporting particles selection (LPS). This algorithm selects surrounding particles within a critical distance of $d_{cri} = 1.5\Delta r$ from a target particle. This is similar to the nearest neighboring particles searching (NNPS) algorithm used in the SPH method.

It is worth noting that the critical distance used in the SPH method is typically larger (around $2.4\Delta r$) for accuracy reasons. As a result, the SPH method typically utilizes approximately 20 supporting particles on average in 2D cases. In contrast, the local L-GSM method uses approximately 8 supporting particles on average, while still ensuring linearly consistent approximation.

6.4.3 *Remarks*

In the SPH method, the nearest neighboring particle searching process is typically conducted at every time step to ensure accuracy. This is because nearby particles may come into the smoothing domain of a particle, and the accuracy of the particle approximation in SPH relies on a good distribution of a sufficiently large number of particles.

On the other hand, the accuracy of both the global and local L-GSM methods is ensured in theory and formulation, regardless of the irregularity of the supporting particles, as long as it is not too extreme, resulting in unphysical approximation. Hence, there is no need to update the particles' connection frequently in the L-GSM method, which makes it much more efficient compared to SPH. Numerical experiments have shown that updating the particles' connection once every 100 time steps is a well-balanced choice between accuracy and efficiency in the L-GSM method [10].

6.5 Construction of smoothing domain

After determining supporting particles for a particle, a gradient smoothing domain or GSD for the particle can then be created for computing smoothed gradients.

To compute the smoothed gradient, a gradient smoothing domain (GSD) is created for each target particle based on the supporting particles.

6.5.1 *For 2D problems*

In a 2D domain, an n-GSD for a target particle can be constructed by successively connecting the midpoints (between the particle and its supporting particles) with the centroids of the triangular cells, as illustrated in Figure 6.6. To do so, the supporting particles must be ordered in sequence in s counterclockwise manner, as shown in Figure 6.6, with respect to the target particle i.

6.5.2 *For 3D problems*

In a 3D domain, the supporting particles for a target particle are first mapped onto a unit's spherical surface, with the target particle at the origin, as shown in Figure 6.7(a). This is achieved by scaling the distance of each

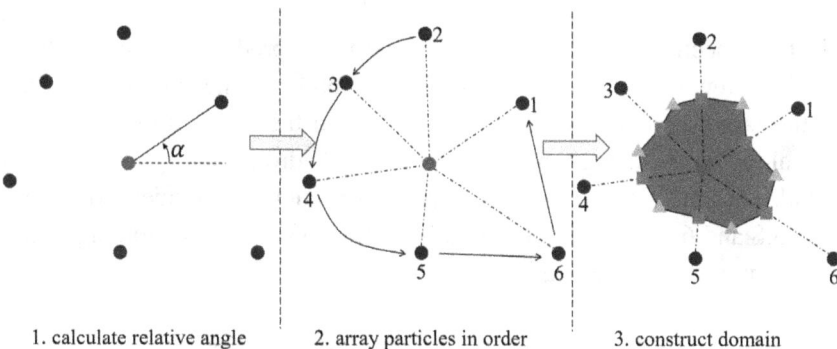

1. calculate relative angle   2. array particles in order   3. construct domain

Figure 6.6. Construction of an nGSD for particle i. (1) particle i and its supporting particles; (2) arrangement of the supporting particles counterclockwise; (3) constructed nGSD.

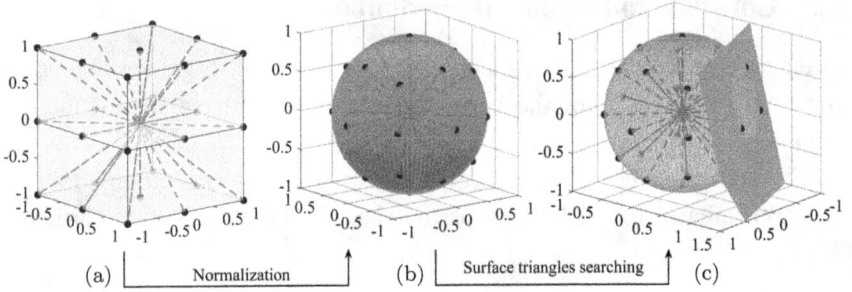

Figure 6.7. Illustration of the "normalization" of supporting particles and the "surface triangle criterion" (Reprinted with permission from [12]).

supporting particle from the target particle to unity, while maintaining the direction.

$$\hat{x}_j = \frac{x_{ij}}{L_{ij}}; \quad \hat{y}_j = \frac{y_{ij}}{L_{ij}}; \quad \hat{z}_j = \frac{z_{ij}}{L_{ij}} \tag{6.40}$$

with

$$L_{ij} = \sqrt{x_{ij}^2 + y_{ij}^2 + z_{ij}^2} \tag{6.41}$$

$$\begin{aligned}
x_{ij} &= x_j - x_i; \\
y_{ij} &= y_j - y_i; \\
z_{ij} &= z_j - z_i.
\end{aligned} \tag{6.42}$$

Next, connecting any three supporting particles produces a unique plane. If all the remaining supporting particles lie on the same side of this plane as the target particle (see Figure 6.7(c)), then the three supporting particles along with the target particle form a tetrahedron. This approach works for all possible distributions of supporting particles, as evidenced in Figures 6.8 and 6.9. The MATLAB code of 3D nGSD construction algorithm is presented and explained in detail in Chapter 9.

6.6 Numerical oscillation and artificial viscosity

It is well known that the SPH method may produce unphysical oscillations in the absence of a viscous term in the governing equations. Similarly, L-GSM can also exhibit such behavior, as shown in Figure 6.10.

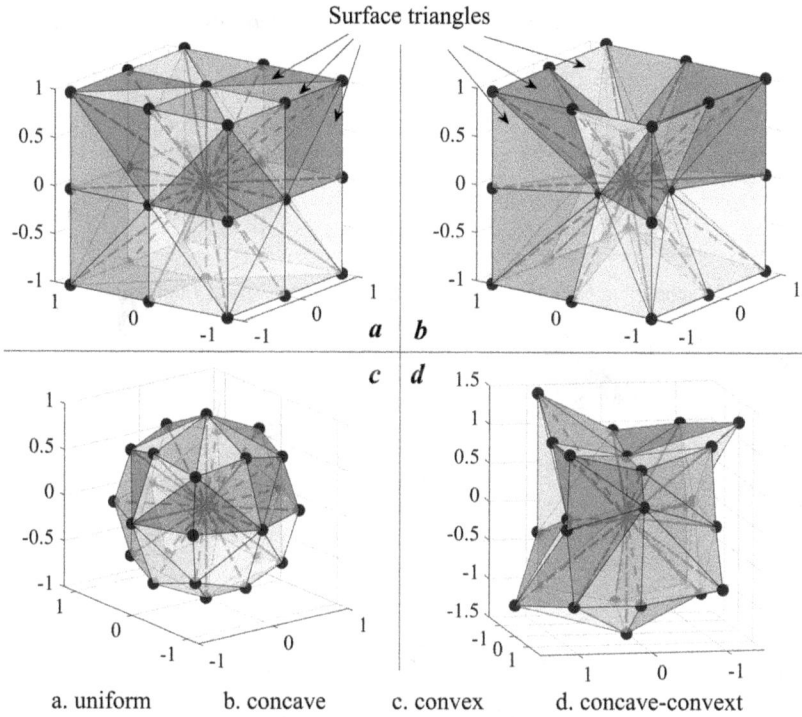

Figure 6.8. Geometry of the polyhedron constructed with the 3D GSD-constructing algorithm for interior particles under various conditions (Reprinted with permission from [12]).

This is because the solution to the original governing equations inherently possesses unstable characteristics, which can be triggered by oscillations in the initial or boundary conditions during a numerical procedure. These characteristics can lead to unstable results, such as shocks and discontinuities that cannot be eliminated without the inclusion of a damping in the momentum equation. Therefore, an artificial viscosity term must be added to the momentum equation to suppress the unphysical oscillations. Monaghan *et al.* [15, 16], Hernquist *et al.* [17], and Lattanzio *et al.* [18] have developed three widely used forms of artificial viscosity.

For hydrodynamics problems, the artificial viscosity term $\mathbf{F}^{viscosity}$ can be obtained by evaluating the viscosity expression for fluid flows [15, 16]:

$$\mathbf{F}^{viscosity} = \left(\frac{1}{\rho}\nabla \cdot \mu\nabla\right)\mathbf{v} \qquad (6.43)$$

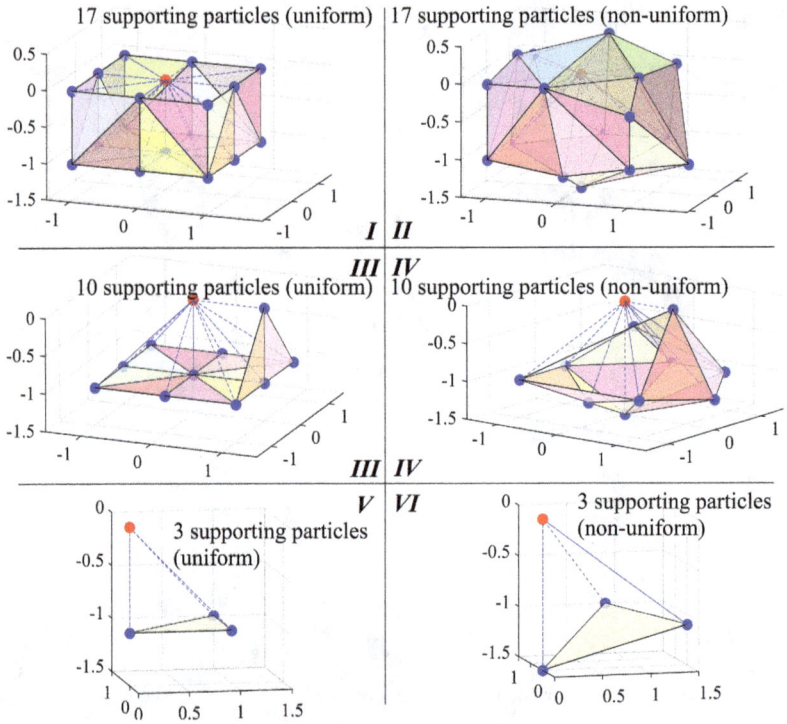

Figure 6.9. Geometry of polyhedron constructed by the 3D GSD-constructing algorithm for boundary particles (Reprinted with permission from [12]).

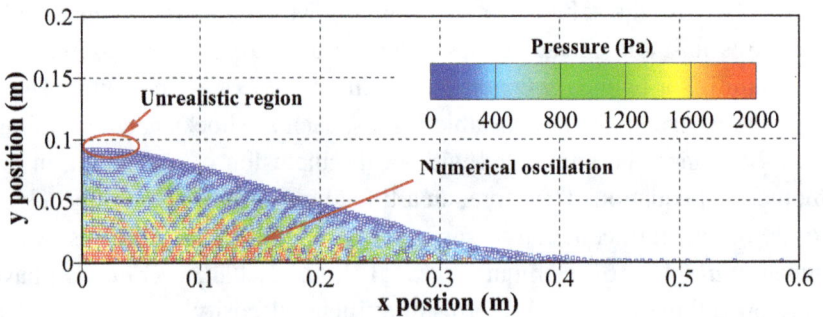

Figure 6.10. Numerical oscillation of pressure in L-GSM simulation of soil column collapse (Reprinted with permission from [19]).

where $\mu = \rho v$ is the dynamic viscosity and v is the kinematic viscosity of the fluid media.

Applying the smoothed gradient to Eq. (6.43) gives the discretized form of hydrodynamics with artificial viscosity:

$$
\begin{cases}
\dfrac{D\rho_i}{Dt} = -\dfrac{\rho_i}{V_i} \displaystyle\sum_{k=1}^{n_i} \left(\dfrac{v_{j_k}^\alpha + v_i^\alpha}{2} \right) (\Delta S_\alpha)_{ij_k} \\[4mm]
\dfrac{Dv_i^\alpha}{Dt} = -\dfrac{1}{\rho_i V_i} \displaystyle\sum_{k=1}^{n_i} \left(\dfrac{p_{j_k} + p_i}{2} \right) (\Delta S_\alpha)_{ij_k} \\[4mm]
\qquad\quad + \dfrac{1}{\rho_i V_i} \displaystyle\sum_{k=1}^{n_i} \dfrac{(\mu_{j_k} + \mu_i) v_{ij_k}^\alpha}{2 r_{ij_k}^2} (x_{ij_k}^\beta (\Delta S_\beta)_{ij_k}) + F^\alpha \\[4mm]
\dfrac{Dd^\alpha}{Dt} = v^\alpha
\end{cases}
\tag{6.44}
$$

where

$$
v_{ij_k}^\alpha = v_i^\alpha - v_{j_k}^\alpha \quad x_{ij_k}^\beta = x_i^\beta - x_{j_k}^\beta \tag{6.45}
$$

The introduction of the artificial viscosity into Eq. (6.44) does not influence the conservation of momentum because of the symmetric feature of the artificial viscosity.

For geomechanics problems, the form of artificial viscosity designed for SPH by Monaghan [15, 20] is adopted in L-GSM. After integrating the artificial viscosity with the smoothed gradient and plugging into the momentum equation, the discretized governing equations have the form of

$$
\begin{cases}
\dfrac{D\rho}{Dt} = -\dfrac{\rho_i}{m_V} \displaystyle\sum_{k=1}^{n_i} \left(\dfrac{v_{j_k}^\alpha + v_i^\alpha}{2} \right) (\Delta S_\alpha)_{ij_k} \\[4mm]
\dfrac{Dv^\alpha}{Dt} = \dfrac{1}{\rho_i V_i} \displaystyle\sum_{k=1}^{n_k} \left(\dfrac{\sigma_{j_k}^{\alpha\beta} + \sigma_i^{\alpha\beta}}{2} - \Pi_{ij_k} \delta^{\alpha\beta} \right) (\Delta S_\alpha)_{ij_k} + F^\alpha \\[4mm]
\dfrac{Dd^\alpha}{Dt} = v^\alpha
\end{cases}
\tag{6.46}
$$

with

$$
\Pi_{ij_k} = \begin{cases} \rho_{ij_k}(-\alpha_\Pi c_s \phi_{ij_k} + \beta_\Pi \phi_{ij_k}^2), & v_{ij_k} \cdot x_{ij_k} < 0 \\ 0, & v_{ij_k} \cdot x_{ij_k} \geq 0 \end{cases} \tag{6.47}
$$

$$\phi_{ij_k} = \frac{\Delta x v_{ij_k}^{\alpha} x_{ij_k}^{\alpha}}{x_{ij_k}^{\alpha} x_{ij_k}^{\alpha} + 0.01 \Delta r^2} \tag{6.48}$$

$$\rho_{ij} = \frac{\rho_i + \rho_j}{2} \tag{6.49}$$

$$x_{ij_k}^{\alpha} = x_i^{\alpha} - x_{j_k}^{\alpha} \quad v_{ij_k}^{\alpha} = v_i^{\alpha} - v_{j_k}^{\alpha} \tag{6.50}$$

where c_s is the sound speed in the media and evaluated by $c_s = \sqrt{E/\rho}$ with E being the young's modulus of the material. α_Π and β_Π are two hyperparameters in the artificial viscosity formulation chosen by the analyst for given problem. For soil problems $\alpha_\Pi = 0.1$, $\beta_\Pi = 0$ is suggested [6].

6.7 On conservation of L-GSM

The L-GSM model assumes constant and identical mass for each particle, allowing automatic satisfaction of mass conservation, as discussed in Section 6.1. However, to achieve momentum conservation, the acceleration of each pair of interacting particles must be anti-symmetric, meaning that $a_{ij} = -a_{ji}$, where a_{ij} is the acceleration of particle i due to the influence of its supporting particle j.

From the momentum equation in Eq. (6.14), we get

$$a_{ij} = \frac{1}{\rho_i V_i} \frac{\sigma_j^{\alpha\beta} + \sigma_i^{\alpha\beta}}{2} (\Delta S_\alpha)_{ij} \tag{6.51}$$

$$a_{ji} = \frac{1}{\rho_j V_j} \frac{\sigma_i^{\alpha\beta} + \sigma_j^{\alpha\beta}}{2} (\Delta S_\alpha)_{ji} \tag{6.52}$$

The conservation of momentum requires the satisfaction of conditions

$$\rho_i V_i = \rho_j V_j \tag{6.53}$$

and

$$(\Delta S_\alpha)_{ij} = -(\Delta S_\alpha)_{ji} \tag{6.54}$$

Condition (6.54) can only be met with a global background mesh as the particles neighbors i and j share a common GSD boundary as shown in Figure 6.11(a). In contrast, the local L-GSM does not usually satisfy the

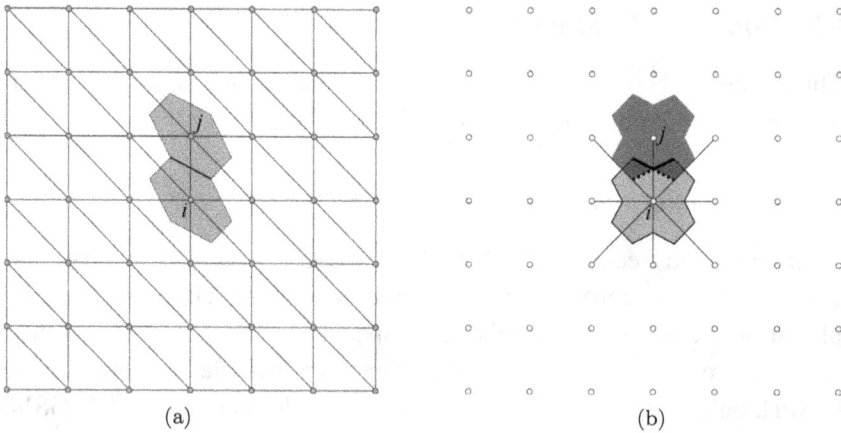

Figure 6.11. Comparison of nGSDin L-GSM. (a) global mesh is used, where the particle pair i and j has a common nGSD boundary segment. (b) local mesh is used, where the nGSDs do not share boundary segments.

condition (6.54) because their nGSD boundaries between each pair of interacting particles are not shared (shown in Figure 6.11(b)).

Thus, the global L-GSM conserves momentum while the local L-GSM does not precisely.

Although the local L-GSM is not precisely conservative, it is still a valuable model for the following reasons:

(1) Local L-GSM is much easier to implement and more efficient to compute without using a global mesh.
(2) Local L-GSM solution is an accurate approximation, and it will converge when the mesh is refined.
(3) Our numerical tests showed that the difference between the L-GSM and local L-GSM solutions is negligible.
(4) Conservation is a good property to have, but it is not necessary to have for a numerical method. As long as the conservation is approximated with good accuracy and the numerical solution converges, the numerical model is valid.

Without insisting on momentum conservation, local GSM can use the real density and area directly, without using the nominal values.

6.8 Boundary treatments

Three types of boundary conditions are introduced: free-slip solid boundary, no-slip solid boundary, and periodic boundary [9].

6.8.1 *Free-slip solid boundary treatments*

To simulate the free-slip solid boundary in L-GSM, fixed virtual particles are employed, as done in SPH (Morris *et al.*, 1997). These particles are placed on the boundary with the same spacing as real particles, and their densities are set equal to those of the nearest real particle. However, unlike in SPH, only a single layer of fixed virtual particles is required in L-GSM. This is because the gradient calculation of field variables in L-GSM only requires the nearest layer of neighboring particles as the supporting particles, as illustrated in Figure 6.12.

To mimic the free-slip effect in the parallel direction on the solid boundary, the velocity of solid boundary particles is set equal to that of the nearest real particle, while the pressure is set to zero. This ensures that the solid boundary particles do not affect the density and stress state of the approaching real particles or influence their moving speed in the tangent direction of the solid boundary edge.

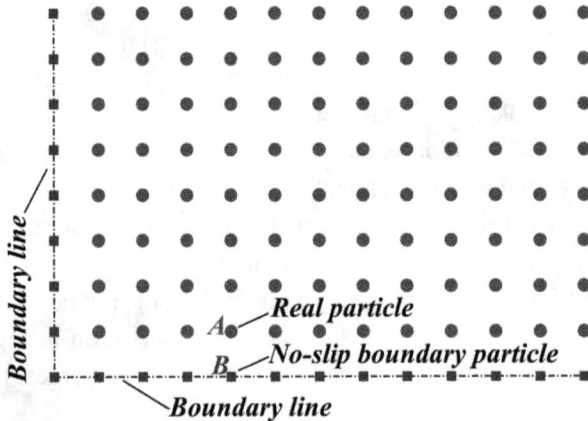

Figure 6.12. Arrangement of boundary particles for free-slip solid boundary (Reprinted with permission from [9]).

To mimic the solid boundary effect in the perpendicular direction, sufficient repulsive force must be provided by the solid boundary particles to prevent real particles from penetrating the solid boundary during simulation. This is achieved by assigning the moving speed of the solid wall to the solid boundary particles in the perpendicular direction, allowing the boundary particles to produce a repulsive pressure/stress. This assignment can provide a low level of repulsive force and works well for slow-flowing problems. However, for high-velocity impact problems, this treatment may not be sufficient to prevent particle penetration. In such cases, the stress tensor of the nearest real particle is assigned to the solid boundary particle, which can provide a higher level of repulsive force to eliminate the unphysical penetration problem.

6.8.2 *No-slip solid boundary treatment*

In the perpendicular direction, the solid boundary effect is simulated in the same way as that in the free-slip boundary treatment. For the parallel direction, the no-slip effect is achieved by assigning the moving speed and pressure of the nearest approaching real particles to the solid boundary particle, which generates an extra density update and repulsive stress or pressure to prevent further relative motion.

Numerical experiments show that the no-slip solid boundary treatment works well for straight boundaries, but not for curved boundaries. Some real particles close to the boundary may experience an unphysical increment of pressure and stress that accumulates over time, repulsing surrounding particles, as shown in Figure 6.13(a), and may eventually cause a blow-up. This phenomenon occurs when the supporting particles of the targeted particle are no longer parallel (see Figure 6.14(a)) to the boundary surface in motion for nonlinear solid boundary.

To overcome this problem, the GSD edges of the real particles connecting with the solid boundary particles should be adjusted to be parallel to the solid boundary locally, as indicated by the "domain edge" in Figure 6.14(b). Computational experiments showed that this treatment can effectively remove the unphysical pressure and stress problem and generate smooth variable fields as shown in Figure 6.13(b) [21].

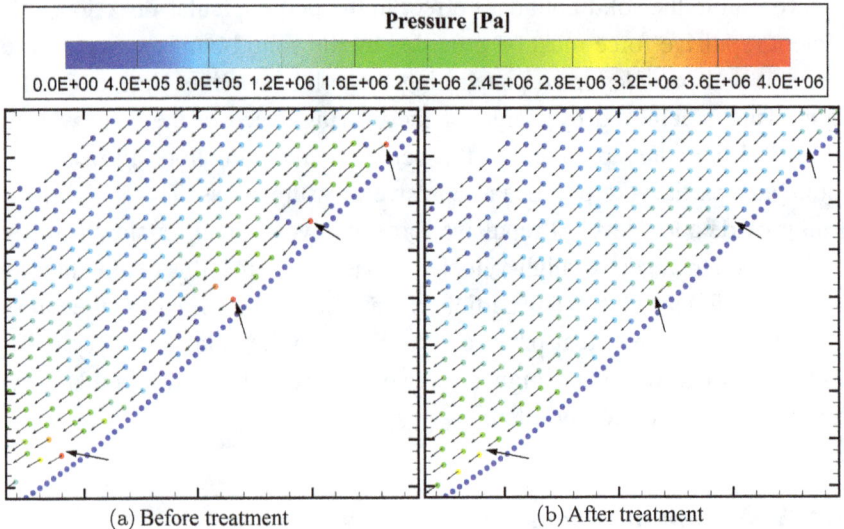

(a) Before treatment　　　　　(b) After treatment

Figure 6.13. Comparison of pressure field of sliding flow before (a) and after (b) the special treatment (Reprinted with permission from [21]).

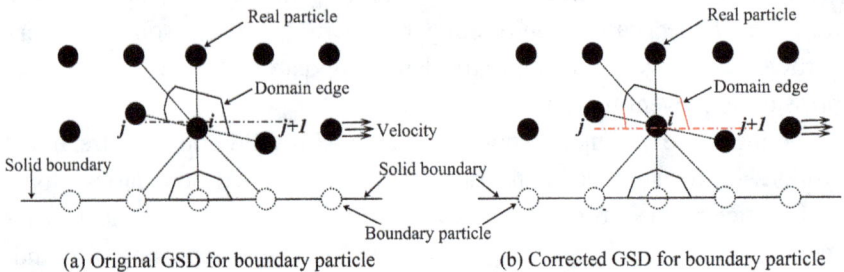

(a) Original GSD for boundary particle　　　　　(b) Corrected GSD for boundary particle

Figure 6.14. GSD adjustment for nonlinear no-slip boundary treatment (Reprinted with permission from [21]).

6.8.3 *Periodical boundary treatment*

To model a periodic boundary condition, the real particles that leave the computational region through a particular boundary face are assumed to immediately reenter the region through the opposite face, with the same variable values except for the x coordinate. The x coordinate is shifted by the length scale L with $x = x - L$, as illustrated in Figure 6.15. For particle interactions, any particle located within the dimension of the GSD from

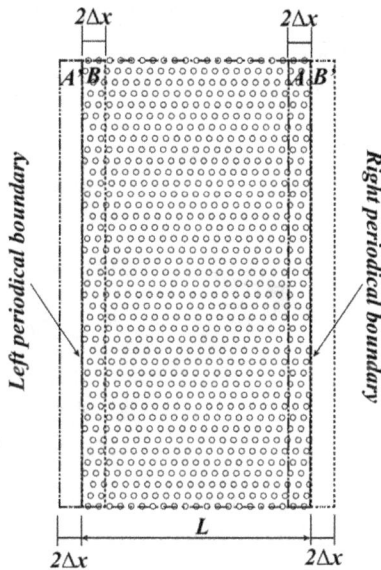

Figure 6.15. Periodical boundary treatment in L-GSM (Reprinted with permission from [9]).

a periodic boundary will interact with particles in an adjacent copy of the system or with particles near the opposite boundary.

In order to implement this procedure, a set of virtual particles is created to mimic the periodic boundary condition, as shown in Figure 6.15. Real fluid particles located in subzone A within a distance of $2\Delta x$ from the left boundary of the computational domain are duplicated directly to subzone A', located next to the right boundary of the domain, by applying $x = x + L$. Similarly, the real fluid particles in subzone B are duplicated to subzone B' by applying $x = x - L$. This wraparound effect of the periodic boundary condition is taken into consideration in both the integration of the equations of motion when moving the particles and in the computation of interactions between supporting particles.

6.9 Time integration

The L-GSM scheme can be integrated with standard methods such as the Leap-Frog (LF), predictor–corrector, and Runge–Kutta schemes, among others. In this study, we adopt the LF scheme considering its lower memory storage and better efficiency.

At each step, the field variables are updated using the explicit LF scheme.

$$
\begin{cases}
\rho_{n+1/2} = \rho_{n-1/2} + \left(\dfrac{D\rho}{Dt}\right)_n \Delta t \\[2ex]
\sigma^{\alpha\beta}_{n+1/2} = \sigma^{\alpha\beta}_{n-1/2} + \left(\dfrac{D\sigma^{\alpha\beta}}{Dt}\right)_n \Delta t \\[2ex]
v^{\alpha}_{n+1/2} = v^{\alpha}_{n-1/2} + \left(\dfrac{Dv^{\alpha}}{Dt}\right)_n \Delta t \\[2ex]
x^{\alpha}_{n+1} = x^{\alpha}_n + v^{\alpha}_{n+1/2}\Delta t
\end{cases}
\tag{6.55}
$$

This scheme is subject to the Courant–Friedrichs–Lewy (CFL) stability condition, which requires the time step to be proportional to the smallest spatial resolution,

$$
\Delta t = \lambda \min\left(\frac{\Delta r}{c_s}\right)
\tag{6.56}
$$

with c_s being the sound speed in the media.

Moreover, due to the inclusion of the artificial viscosity term, by referring to the work for SPH [20, 22], the critical time step must satisfy a more rigid criterion shown in Eq. (6.57) instead of Eq. (6.56). If the material is subject to external force, the critical time step should satisfy Eq. (6.58) [20, 22]

$$
\Delta t_{cv} = \min\left(\frac{\Delta r_i}{1.2(\alpha_\Pi c_{s_i} + \beta_\Pi \max(\phi_{ij}))}\right)
\tag{6.57}
$$

$$
\Delta t_f = \min\left(\frac{\Delta r_i}{f_i}\right)^{\frac{1}{2}}
\tag{6.58}
$$

$$
\Delta t = \lambda \min(\Delta t_{cv}, \Delta t_f)
\tag{6.59}
$$

where α_Π and β_Π are hyper-parameters in the artificial viscosity term shown in Eq. (6.47), and f is acceleration due to external force. To ensure stability, safety factor $\lambda = 0.25$ is used.

The entire calculation process of the L-GSM framework can be shown by the following flowchart in Figure 6.16.

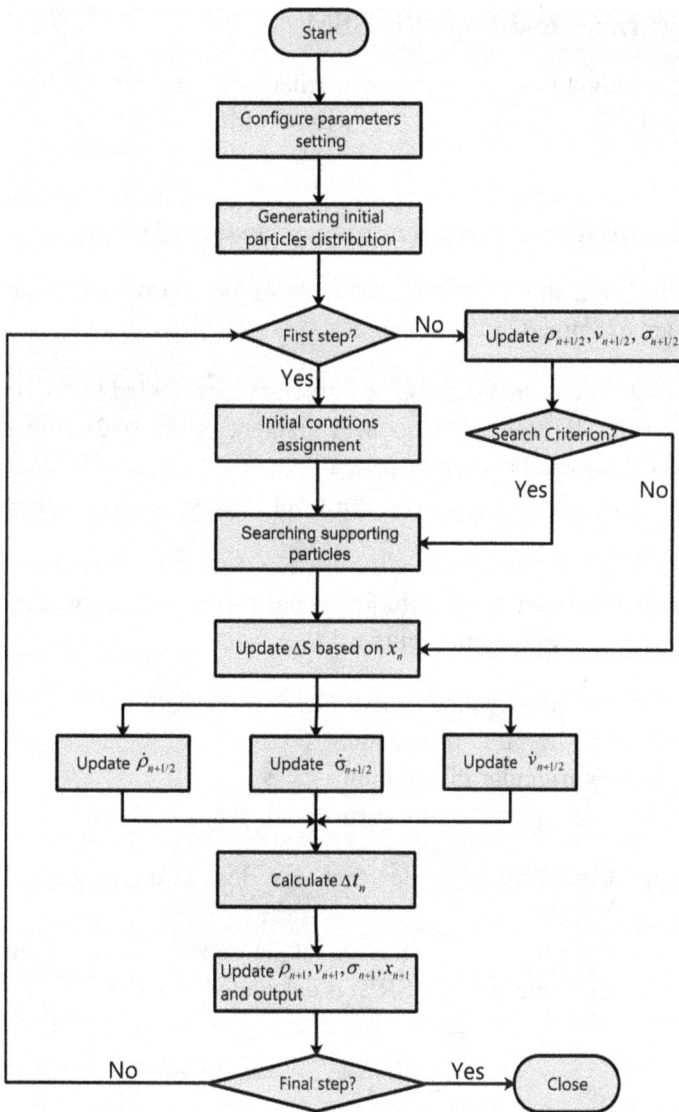

Figure 6.16. Flow chart of the L-GSM solver (Reprinted with permission from [12]).

6.10   Accuracy condition of L-GSM

The accuracy condition of the adopted smoothed gradient has been analyzed theoretically in Section 6.2.3. It will be verified by a series of numerical tests.

6.10.1   *Numerical tests on accuracy of smoothed gradient*

In this subsection, the accuracy conditions of **three forms** of smoothed gradients are examined in the 2D case:

- *Approach I*: The standard smoothed gradient represented by Eq. (6.2).
- *Approach II*: The standard smoothed gradient in Eq. (6.2) with virtual particle assignment beyond boundary.
- *Approach III*: The normalized form of the smoothed gradient in Eq. (6.35).

The accuracy conditions are studied under four different particle distribution scenarios, as shown in Figure 6.17:

- *Case 1*: Uniform distribution of particles.
- *Case 2*: Slightly irregular distribution.
- *Case 3*: Highly irregular distribution.
- *Case 4*: Boundary particle with particle deficiency problem.

The **testing functions** are the first six terms of monomials: 1, x, x^2, x^3, x^4, x^5.

To examine the accuracy condition of numerical solutions, the root mean squared relative error (RMSRE) is calculated as follows:

$$\text{RMSRE} = \sqrt{\frac{\sum_{k=1}^{N} \left(|U_x^{num} - U_x^{exact}| / |U_x^{exact}| \right)_k^2}{N}} \tag{6.60}$$

The numerical testing results in Figure 6.18 and Table 6.1 show that, for inner particles in Cases 1–3, all three schemes provided the same numerical results, which are 2nd order accurate when particles are uniformly distributed in Case 1, and 1st order accurate when particles are irregularly distributed in Cases 2 and 3. The quadratic term x^2 is resolved exactly

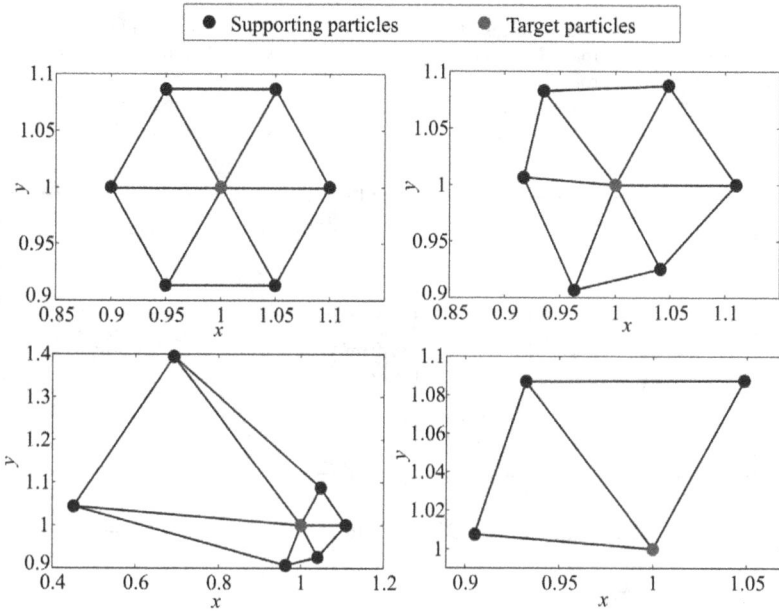

Figure 6.17. Four study cases representing different supporting particles distributions conditions (Reprinted with permission from [9]).

using these three schemes in Case 1, owing to the 2nd-order accuracy. Particularly, approach I is not consistent for boundary particles in Case 4 due to the particle deficiency problem, while approach II and approach III are both still 1st order accurate, as expected. Notably, the exact solutions for constant and linear functions are obtained using any approach, and thus are not presented in the results.

The numerical experiments confirm the conclusions from the theoretical analysis.

6.10.2 *Accuracy comparison of L-GSM and SPH gradient operators*

The compared SPH gradient operator has a form of

$$
\left\langle \frac{\partial U_i}{\partial x_i} \right\rangle = -\sum_{j_k=1}^{N} \frac{m_j}{\rho_{j_k}} U_{j_k} \nabla W(x_i - x_{j_k}, h) = -\sum_{j_k=1}^{N} V_j U_{j_k} \nabla_x W_{ij} \qquad (6.61)
$$

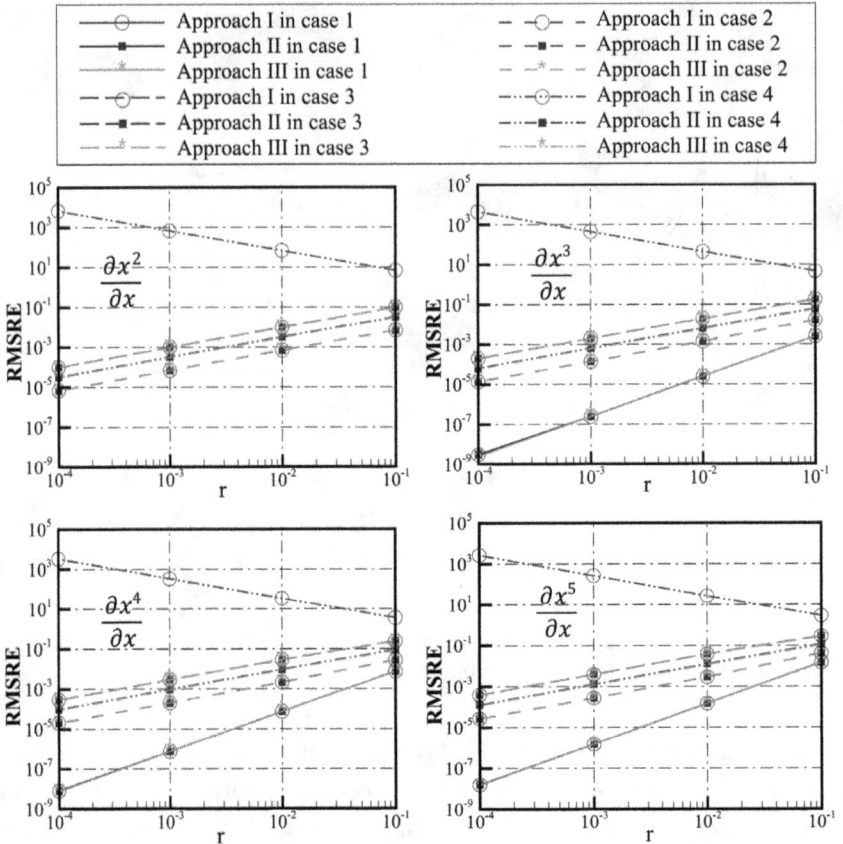

Figure 6.18. RMSRE errors of numerical solutions using the three L-GSM approaches within four cases as shown in Figure 6.17 for the gradient calculation of function x^2, x^3, x^4, x^5, respectively (Reprinted with permission from [9]).

where N is the total number of supporting particles of particle i, m_j, and ρ_j are the mass and density, respectively, represented by the supporting particle j_k. h is called smoothing length related to the size of supporting domains in SPH, which is typically set as $1.2\Delta x$ through balancing the accuracy of numerical solution and the computational efficiency. W is the kernel function or shape function for SPH gradient approximation, and

Table 6.1. Gradient approximation results using the three L-GSM schemes within the four different cases shown in Figure 6.17 associated with a characteristic length r of 0.01. The target particle is located at $(1, 1)$ [9].

		$\partial(1)/\partial x$	$\partial(x)/\partial x$	$\partial(x^2)/\partial x$	$\partial(x^3)/\partial x$	$\partial(x^4)/\partial x$
Case 1	Approach I	0.0000	1.0000	2.0000	3.0001	4.0003
	Approach II	0.0000	1.0000	2.0000	3.0001	4.0003
	Approach III	0.0000	1.0000	2.0000	3.0001	4.0003
Case 2	Approach I	0.0000	1.0000	2.0014	3.0042	4.0085
	Approach II	0.0000	1.0000	2.0014	3.0042	4.0085
	Approach III	0.0000	1.0000	2.0014	3.0042	4.0085
Case 3	Approach I	0.0000	1.0000	1.9836	2.9514	3.9038
	Approach II	0.0000	1.0000	1.9836	2.9514	3.9038
	Approach III	0.0000	1.0000	1.9836	2.9514	3.9038
Case 4	Approach I	0.0000	-133.429	-132.574	-131.724	-130.88
	Approach II	0.0000	1.0000	1.9939	2.9820	3.9642
	Approach III	0.0000	1.0000	1.9940	2.9822	3.9645

the widely used cubic B-spline kernel function is applied here

$$
W_{ij_k} = \alpha_d \times
\begin{cases}
\dfrac{2}{3} - q^2 + \dfrac{1}{2}q^3, & 0 \le q < 1 \\[2mm]
\dfrac{1}{6}(2-q)^3, & 1 \le q < 2 \\[2mm]
0, & q \ge 2
\end{cases}
\tag{6.62}
$$

where $\alpha_d = 15/(7\pi h^2)$ for the 2D problem and $\alpha_d = 3/(2\pi h^3)$ for 3D problem from the normalization condition, and q is defined as the relative distance between the particles i and j_k by $q = r/h$, where r stands for the distance between two particles.

The accuracy condition of the SPH operator is compared to the smoothed gradient in global L-GSM and local L-GSM by approximating the gradient of function $f = x^2 + y^2$. For better representation, irregularly distributed particles are employed as shown in Figure 6.19.

The results in Figure 6.20 show that all three gradient operators are 1st order accurate as expected with the relative error of the smoothed gradient

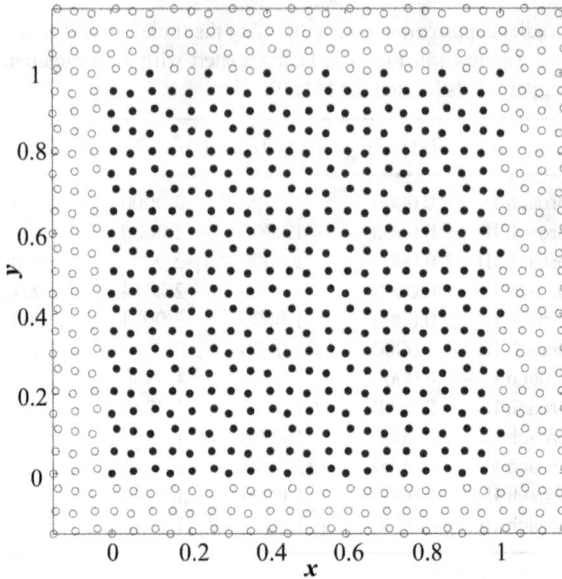

Figure 6.19. Computational domain and irregular particle distribution with $r = 0.05$. The dots are "real" particles within the computational domain included in the gradient calculation while the boundary particles shown in circles (at least 3 layers for SPH) are assigned to support the real particles near the boundary of the domain.

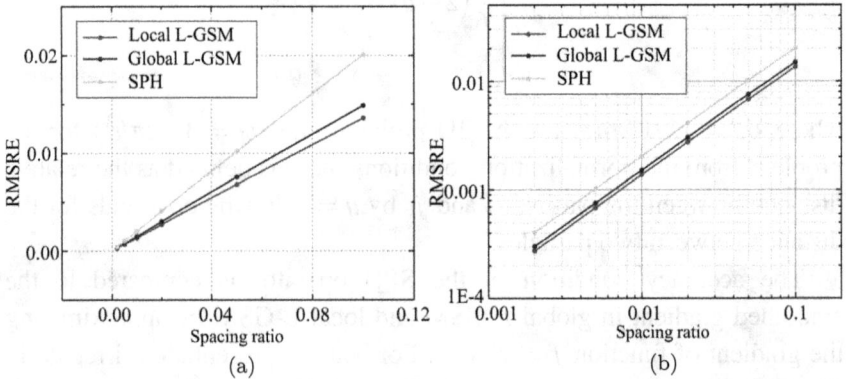

Figure 6.20. Comparison RMSRE errors of local L-GSM, global L-GSM, and SPH in both linear scale (a) and log scale (b) (Figure adapted with permission from [10]).

Table 6.2. Average number of neighboring or supporting particles used in the gradient approximation [10].

Schemes	Local L-GSM	Global L-GSM	SPH
Ave. # of Neighbors	7.54	5.99	18.67

in local L-GSM being the smallest and SPH showing the largest error. The local L-GSM is found more accurate than the global L-GSM due to its use of a larger number of supporting particles in gradient approximation as listed in Table 6.2.

6.10.3 *Accuracy of 3D smoothed gradient in local L-GSM*

The accuracy of the 3D smoothed gradient in Eq. (6.12) is compared to the normalized 3D SPH gradient approximation technique. The normalized form of 3D SPH gradient formulation is

$$
\begin{cases}
\left.\dfrac{\partial f}{\partial x}\right|_i = \dfrac{\displaystyle\sum_{j=1}^{n_i}(f_j - f_i)dW_{ij}dx_j}{\displaystyle\sum_{j=1}^{n_i}(x_j - x_i)dW_{ij}dx_j} \\[6mm]
\left.\dfrac{\partial f}{\partial y}\right|_i = \dfrac{\displaystyle\sum_{j=1}^{n_i}(f_j - f_i)dW_{ij}dy_j}{\displaystyle\sum_{j=1}^{n_i}(y_j - y_i)dW_{ij}dy_j} \\[6mm]
\left.\dfrac{\partial f}{\partial z}\right|_i = \dfrac{\displaystyle\sum_{j=1}^{n_i}(f_j - f_i)dW_{ij}dz_j}{\displaystyle\sum_{j=1}^{n_i}(z_j - z_i)dW_{ij}dz_j}
\end{cases}
\tag{6.63}
$$

The normalized SPH gradient approximation technique can restore the 1st-order accuracy even for the boundary particles with particle deficiency.

Table 6.3. Percentage (%) of RMSRE errors of 3D local L-GSM and SPH solutions to approximating the gradient of Eq. (6.64) when particles are distributed uniformly [12].

	Local L-GSM			SPH		
r	$\partial f/\partial x$	$\partial f/\partial y$	$\partial f/\partial z$	$\partial f/\partial x$	$\partial f/\partial y$	$\partial f/\partial z$
0.2	2.20E-14	0.129204	0.192721	1.02E-13	0.161548	0.240965
0.1	5.00E-14	0.032332	0.04836	1.07E-13	0.040426	0.060466
0.05	8.40E-14	8.09E-03	1.21E-02	8.9E-14	0.010111	0.015145
0.02	1.58E-13	1.29E-03	1.94E-03	1.58E-13	0.001618	0.002426
0.01	2.78E-13	3.24E-04	4.85E-04	2.7E-13	0.000405	0.000607

Table 6.4. Percentage (%) of RMSRE errors of 3D local L-GSM and SPH solutions to approximating the gradient of Eq. (6.64) when particles are distributed irregularly [12].

	Local L-GSM			SPH		
r	$\partial f/\partial x$	$\partial f/\partial y$	$\partial f/\partial z$	$\partial f/\partial x$	$\partial f/\partial y$	$\partial f/\partial z$
0.2	1.16E-01	0.445249	0.374797	0.149372	0.573926	0.483113
0.1	5.80E-02	0.222624	0.187398	0.074767	0.286963	0.241556
0.05	2.90E-02	0.111312	0.093699	0.037372	0.143481	0.120778
0.02	1.16E-02	4.47E-02	3.76E-02	0.014942	0.057623	0.048505
0.01	5.80E-03	2.24E-02	1.88E-02	0.00747	0.028812	0.024253

The gradient of the following polynomial function is approximated with these two methods for the domain of unit cube.

$$f(x,y,z) = \frac{x^2}{2} + \frac{y^3}{3} + \frac{z^4}{4} + 10(x+y+z) \tag{6.64}$$

The accuracy conditions of the 3D smoothed gradient by local L-GSM and SPH under different resolutions $r = 0.2/0.1/0.05/0.02/0.01$ are listed in Tables 6.3 and 6.4. The results are also plotted in Figure 6.21 for the uniformly and irregularly distributed particles. It is found that (i) both approaches are 2nd order accurate when particles are distributed uniformly and 1st order accurate when distributed irregularly, (ii) the 3D smoothed gradient has a relatively smaller error than the normalized 3D SPH gradient technique.

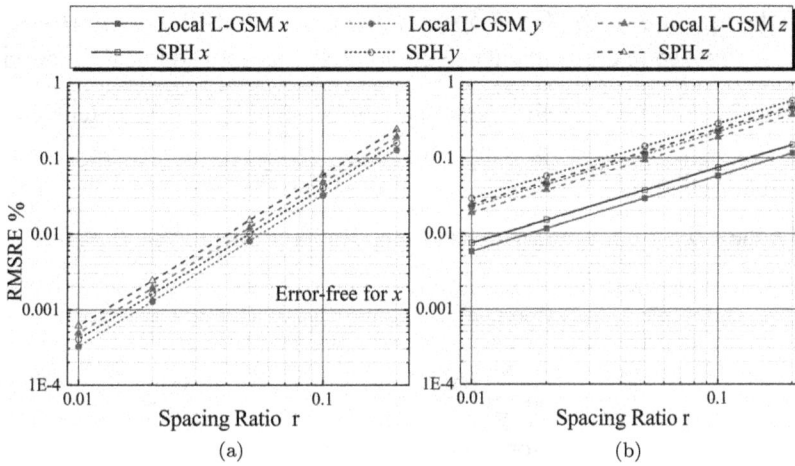

Figure 6.21. Comparison of RMSRE percentages of 3D local L-GSM and SPH solutions to approximating the gradient of Eq. (6.64) when particles are distributed uniformly (a) and irregularly (b). In the legend, x, y, and z represent, $\partial f/\partial x$, $\partial f/\partial y$, and $\partial f/\partial z$, respectively (Figure adapted with permission from [12]).

6.10.4 *Remarks*

Based on theoretical studies and numerical experiments, the following conclusions can be drawn:

- For inner particles, the standard smoothed gradient formulations in Eqs. (6.2) and (6.12) are 1st order accurate in general cases and 2nd order accurate when particles are distributed uniformly.
- For boundary particles, the standard smoothed gradient formulations in Eqs. (6.2) and (6.12) are inaccurate due to "particles deficiency".
- The "virtual particle" treatment and the normalized smoothed gradient presented in Section 6.3 are proven effective in holding the accuracy condition of the smoothed gradient for boundary particles.
- The relative error of the GSM smoothed gradient is always slightly smaller than that of the SPH gradient technique under the same conditions because of more geometry information employed by the GSM operator.
- The smoothed gradient in local L-GSM is slightly more accurate than that in L-GSM, due to more supporting particles used in the gradient approximation.

- A key feature of the GSM operator, compared to SPH, is its error-free performance in the gradient calculation for constant and linear functions.

In conclusion, the smoothed gradient can be regarded as the extension of the 2nd-order accurate central finite difference scheme for the cases of unstructured grids.

Take the 2D formulation as an example. The proof is presented as follows.

The smoothed gradient of function U at point i in terms of x can be written as

$$\frac{\partial U_i}{\partial x} = \frac{U_i}{2V_i} \sum_{k=1}^{n_i} (\Delta S_x)_{ij_k} + \frac{1}{2V_i} \sum_{k=1}^{n_i} (\Delta S_x)_{ij_k} U_j \tag{6.65}$$

According to the deduction works regarding $A_1 = 0$ in Eq. (6.20), it is obvious that the first term on the right-hand side of Eq. (6.65) vanishes.

In the cases where particles are distributed uniformly, if we denote the points using the row number and column number regularly as done in FDM, then the supporting particles for a given particle $P_{I,J}$ at the Ith row and Jth column will be $P_{I-1,J}, P_{I+1,J}, P_{I,J-1}$, and $P_{I,J+1}$. Hence,

$$\frac{\partial U_i}{\partial x} = \frac{1}{2V_i} \left((\Delta S_x)_{I+1,J} U_{I+1,J} + (\Delta S_x)_{I,J+1} U_{I,J+1} \right.$$
$$\left. + (\Delta S_x)_{I-1,J} U_{I-1,J} + (\Delta S_x)_{I,J-1} U_{I,J-1} \right) \tag{6.66}$$

Using Eq. (6.4), we have

$$(\Delta S_x)_{I+1,J} = \frac{y_{I,J+1} - y_{I,J-1}}{3} = \frac{2h}{3}; \quad (\Delta S_x)_{I,J+1} = \frac{y_{I-1,J} - y_{I+1,J}}{3} = 0;$$

$$(\Delta S_x)_{I-1,J} = \frac{y_{I,J-1} - y_{I,J+1}}{3} = -\frac{2h}{3}; \quad (\Delta S_x)_{I,J-1} = \frac{y_{I+1,J} - y_{I-1,J}}{3} = 0; \tag{6.67}$$

where h denotes the spacing ratio of the uniformly distributed particles.

Since the area V_i of the smoothing domain for the given particle i (or $P_{I,J}$) is calculated as $2/3h^2$ by doing some simple algebraic work, substituting the area V_i and the shape functions in Eq. (6.67) into Eq. (6.66) yields

$$\frac{\partial U_i}{\partial x} = \frac{U_{I+1,J} - U_{I-1,J}}{2h} \tag{6.68}$$

The right-hand side of Eq. (6.68) is exactly the central finite difference scheme of $\partial U_i/\partial x$ in the 2D case. That is, the GSM smoothed gradient is equivalent to the central finite difference scheme as particles are distributed uniformly in the 2D case. Similar analysis can be done in the 3D case, which will give the same conclusion.

6.11   Stability condition of L-GSM

The SPH method has been known to have a "tensile instability" problem [6–8], which affects its stability and accuracy. To address this issue, the stability condition of the L-GSM method has been theoretically analyzed through von Neumann stability analysis [9], and it has been shown that L-GSM is free from the tensile instability issue, making it more advantageous than SPH.

To verify this, a simple numerical example is conducted to examine the tensile stability condition of 2D L-GSM. Two layers of boundary particles are fixed out of interior particles to avoid wave propagation, and the interior particles are free to move under external perturbations. The regularly or irregularly distributed particles are initially under uniform tensile

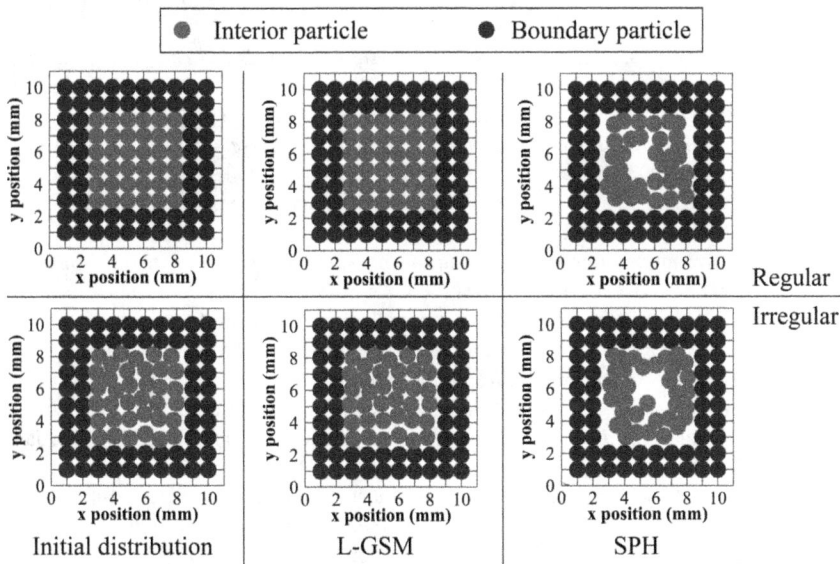

Figure 6.22.   Schematic illustration of the particle distribution at initial state (left), and at 100 μs using L-GSM (Middle) and SPH (right) method (Reprinted with permission from) [9].

stress. If one applies a very small velocity perturbation, say 10^{-10}km/s, to a center particle, theoretically, it would take 10^4 s for a particle to move a distance equal to the particle separation. However, the result shown in Figure 6.22 is obtained at only 100 μs. The numerical results demonstrate that the SPH method suffers from the tensile instability while the L-GSM scheme is stable in tension, regardless of whether the initial particle distribution is regular or irregular.

In the next chapter, we will apply the L-GSM to solve problems of incompressible hydrodynamics.

References

[1] G. R. Liu and M. B. Liu, *Smoothed Particle Hydrodynamics: A Meshfree Particle Method.* Singapore: World Scientific Publishing Co., (2003).

[2] M. B. Liu and G. R. Liu, Smoothed Particle Hydrodynamics (SPH): An Overview and Recent Developments, *Archives of Computational Methods in Engineering*, 17(1), 25–76 (2010, March).

[3] G. R. Liu, "An Overview on Meshfree Methods: For Computational Solid Mechanics," *International Journal of Computational Methods*, 13(5), 1630001 (2016, October). doi: 10.1142/S0219876216300014.

[4] G. R. Liu and Y. T. Gu, *An Introduction to Meshfree Methods and Their Programming*, 1st ed. Netherlands: Springer Netherlands, (2005). Accessed: (2019, November 9) [Available: https://www.springer.com/gp/book/9781402032288.

[5] G. R. Liu, *Finite Element Method: A Practical Course.* Oxford: Butterworth-Heinemann (2003).

[6] H. H. Bui, R. Fukagawa, K. Sako, and S. Ohno, Lagrangian meshfree particles method (SPH) for large deformation and failure flows of geomaterial using elastic-plastic soil constitutive model, *International Journal for Numerical and Analytical Methods in Geomechanics*, 32(12), 1537–1570 (2008, August).

[7] Z. Mao, G. R. Liu, and X. Dong, A comprehensive study on the parameters setting in smoothed particle hydrodynamics (SPH) method applied to hydrodynamics problems, *Computers and Geotechnics*, 92, 77–95 (2017).

[8] J. W. Swegle, D. L. Hicks, and S. W. Attaway, Smoothed Particle Hydrodynamics Stability Analysis, *Journal of Computational Physics*, 116(1), 123–134 (1995, January).

[9] Z. Mao and G. R. Liu, A Lagrangian gradient smoothing method for solid-flow problems using simplicial mesh, *International Journal for Numerical Methods in Engineering*, 113(5), 858–890 (2018, February). doi: 10.1002/nme.5639.

[10] Z. Mao, G. R. Liu, and Y. Huang, A local Lagrangian gradient smoothing method for fluids and fluid-like solids: A novel particle-like method, *Engineering Analysis with Boundary Elements*, 107, 96–114 (2019, October). doi: 10.1016/j.enganabound.2019.07.003.

[11] Z. Mao and GR. Liu, A 3D L-GSM framework with an adaptable GSD-constructing algorithm for simulating large deformation free surface flows, *International Journal for Numerical Methods in Engineering*, 6265 (2019, October). doi: 10.1002/nme.6265.

[12] Z. Mao and G. R. Liu, A 3D Lagrangian gradient smoothing method framework with an adaptable gradient smoothing domain-constructing algorithm for simulating large deformation free surface flows, *International Journal for Numerical Methods in Engineering*, 121(6), 1268–1296 (2020, March). doi: 10.1002/nme.6265.

[13] D. T. Lee and B. J. Schachter, Two algorithms for constructing a Delaunay triangulation, *International Journal of Computer & Information Sciences*, 9(3), 219–242 (1980).

[14] S. Sloan, "A fast algorithm for constructing Delaunay triangulations in the plane," *Advances in Engineering Software*, 9(1), 34–55 (1987).

[15] J. J. Monaghan, artificial viscosity for particle methods J.J. Monaghan and H. Pongracic, *Applied Numerical Mathematics*, 1, 187–194 (1985).

[16] J. J. Monaghan, Simulating Free Surface Flows with SPH, *Journal of Computational Physics*, 110(2), 399–406 (1994, February).

[17] L. Hernquist and N. Katz, "TREESPH — A unification of SPH with the hierarchical tree method," *Astrophysical Journal, Supplement Series*, 70, 419–446 (1989, June).

[18] J. C. Lattanzio, J. J. Monaghan, H. Pongracic, and M. P. Schwarz, "Controlling Penetration," *SIAM Journal on Scientific and Statistical Computing*, 7(2), 591–598 (1986, April).

[19] Z. Mao, A Novel Lagrangian Gradient Smoothing Method for Fluids and Flowing Solids, University of Cincinnati, (2019). Accessed: January 04, 2021. Available: http://rave.ohiolink.edu/etdc/view?acc_num=ucin1553252214052311.

[20] J. J. Monaghan, Smoothed Particle Hydrodynamics, *Annual Review of Astronomy and Astrophysics*, 30(1), 543–574 (1992).

[21] Z. Mao, G. Liu, Y. Huang, and Y. Bao, A conservative and consistent Lagrangian gradient smoothing method for earthquake-induced landslide simulation, *Engineering Geology*, 260 (2019, October). doi: 10.1016/j.enggeo.2019.105226.

[22] J. J. Monaghan, "On the problem of penetration in particle methods," *Journal of Computational Physics*, 82(10), 1–15 (1989).

[23] J. P. Morris, P. J. Fox, and Y. Zhu, Modeling low reynolds number incompressible flows using SPH, *Journal of Computational Physics*, 136(1), 214–226 (1997). https://doi.org/10.1006/jcph.1997.5776.

Chapter 7

L-GSM for Incompressible Hydrodynamics

Contents

This chapter applies global and local L-GSM methods to solve hydrodynamics problems governed by Navier–Stokes equations in PDE form. The efficacy of these techniques is examined through an intensive study of a number of practical engineering problems. The chapter is organized into two sections:

- Section 7.1 discusses the governing equations and equation of state for simulating hydrodynamics problems.
- Section 7.2 presents applications of global and local L-GSM to various problems, including the following:
  - ○ Couette flow
  - ○ Poiseuille flow

- o Shear-driven cavity
- o Dam break
- o Wall impact of breaking dam
- o Water discharge from container
- o Droplet splash

7.1 Governing equations

Assume the fluid media is *inviscid* and *incompressible* or *weakly compressible*. Following the governing PDEs and gradient approximation formulations given in Chapter 6, the corresponding N–S PDEs can have the following discretized form:

$$
\begin{cases}
\dfrac{D\rho_i}{Dt} = -\dfrac{\rho_i}{V_i} \displaystyle\sum_{k=1}^{n_i} \left(\dfrac{v_{j_k}^\alpha + v_i^\alpha}{2} \right) (\Delta S_\alpha)_{ij_k} \\[2ex]
\dfrac{Dv_i^\alpha}{Dt} = -\dfrac{1}{\rho_i V_i} \displaystyle\sum_{k=1}^{n_k} \left(\dfrac{p_{j_k} + p_i}{2} \right) (\Delta S_\alpha)_{ij_k} \\[2ex]
\quad + \dfrac{1}{m_i} \displaystyle\sum_{k=1}^{n_i} \dfrac{(\mu_{j_k} + \mu_i) v_{ij_k}^\alpha}{2 r_{ij_k}^2} \left(x_{ij_k}^\beta (\Delta S_\beta)_{ij_k} \right) + F^{external} \\[2ex]
\dfrac{Dd^\alpha}{Dt} = v^\alpha
\end{cases}
\tag{7.1}
$$

The pressure of particles is density dependent and can be determined by an equation of state (EoS). Two widely used EoS in SPH are adopted in L-GSM.

The first equation [1] predicts pressure from density ρ and the speed of sound c_s:

$$
p = c_s^2 \rho
\tag{7.2}
$$

This EoS works well for low *Reynolds* number flows, like the Couette flows, Poiseuille flow, and shear driven cavity.

One can note that the speed of sound must be chosen carefully to ensure both the efficiency in computation and the accuracy of the numerical solution. This is because a too small c_s value will fail to mimic the incompressible property of flow, while a too large c_s value leads to a smaller critical time step, which will influence the computational efficiency. The selection of c_s in the L-GSM follows the same rule derived for the SPH method:

$$c_s^2 \approx \frac{V_0^2}{\delta}, \quad \frac{vV_0}{L_0\delta}, \quad \frac{FL_0}{\delta} \qquad (7.3)$$

where v is the kinematic viscosity of fluid, F is the body force per unit mass, V_0 and L_0 are the velocity scale and characteristic length scale respectively, and

$$\delta = \frac{\Delta\rho}{\rho_0} \qquad (7.4)$$

with ρ_0 being the reference density or initial density of media.

The other widely used EoS in SPH [2] determines the pressure p with

$$p(\rho) = B\left[\left(\frac{\rho}{\rho_0}\right)^\gamma - 1\right] \qquad (7.5)$$

This equation is used for free surface water flows, including dam breaking, water discharge, and droplet splash [3].

The parameters B and γ are chosen such that the maximum density oscillations remain lower than 1% around the initial density ρ_0. In practice, this is accomplished, following [2], by setting $\gamma = 7$, and

$$B = \frac{200gH}{\rho\gamma} \qquad (7.6)$$

where H is the dam height.

7.2 Numerical examples

7.2.1 Couette flow

In this study, the Couette flow between two infinite plates was modeled using 21×41 particles, including 39 layers of real particles and 2 layers of virtual particles, as shown in Figure 7.1. The flow is initially at rest at time $t = 0$. The upper plate moves at constant velocity V_0 parallel to the x-axis. An analytical solution in the form of series solution for the horizontal velocity field was given by [1]:

$$\mathbf{v}_x(y,t) = \frac{V_0}{L}y + \sum_{n=1}^{\infty} \frac{2V_0}{n\pi}(-1)^n \sin\left(\frac{n\pi}{L}y\right)\exp\left(-v\frac{n^2\pi^2}{L^2}t\right) \qquad (7.7)$$

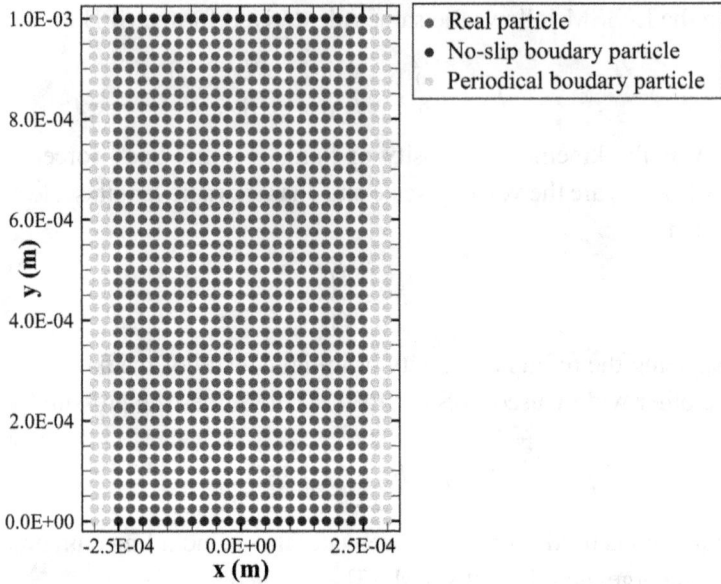

Figure 7.1. Initial geometry and particle distribution for the Couette flow (Reprinted with permission from [4]).

Table 7.1. Parameter settings for the Couette flow (the mass is per particle) [4].

Δx [m]	L [m]	ρ [kg/m³]	Mass [kg]	v [m²/s]	V_0 [m/s]	c_s [m/s]	Δt [s]
2.5×10^{-5}	10^{-3}	10^3	6.25×10^{-7}	10^{-6}	2.5×10^{-5}	10^{-4}	10^{-4}

where v is the kinetic viscosity and L is the width of the flow channel in the y direction.

Numerical simulations are performed using three approaches: global L-GSM, local L-GSM, and SPH. The governing equation (7.1), together with Eq. (7.2), is solved for field variables numerically. A periodic boundary condition presented in Section 6.8.3 is implemented on the left and right boundaries to simulate the infinite flow. With the parameters listed in Table 7.1, the flow reaches a steady state at $t = 20$. The particle distribution with velocity vector at the steady state is plotted in Figure 7.2.

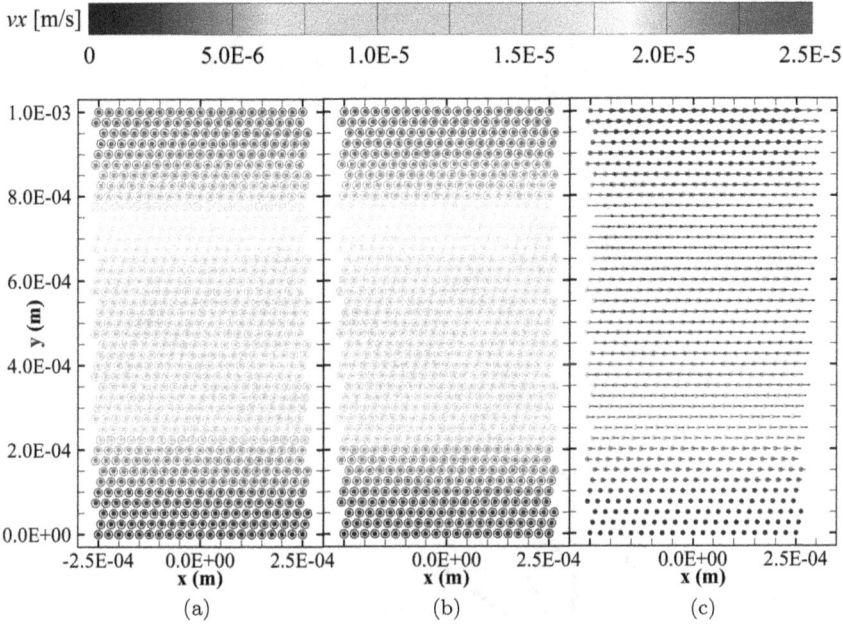

Figure 7.2. Particle distribution comparison in the Couette flow at $t = 1.0$. Local L-GSM (dots), global L-GSM (circles in a), and SPH (circles in b). The velocity quiver by local L-GSM is shown in (c) (Reprinted with permission from [5]).

The accuracy of the numerical solutions is evaluated by comparison with the analytical solution, as shown in Figure 7.3. The velocity values on the center vertical line obtained from the local L-GSM, global L-GSM, and SPH are compared with the analytical solution at $t = 0.01$, 0.1, and the final steady state $t = 20$. The relative error of the numerical solutions from the exact solution is calculated and compared in Figure 7.4 using different particle spacings: $\Delta x = 5.0/2.5/2.0/1.25/1.0/0.5 \times 10^{-5}$. Less than 0.5% discrepancy with the analytical series solution is found as shown in Table 7.2, which validates the L-GSM. As expected, the local L-GSM gives relatively better accuracy than the global L-GSM. This is because more supporting particles are used in the gradient approximation.

Figure 7.5 shows the density variation of the flow generated by the three numerical methods. The variation is smaller than 1%, which verifies the effectiveness of the EoS (7.2) in representing this incompressible flow.

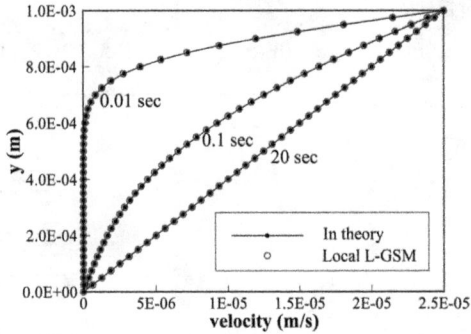

(a) Local L-GSM

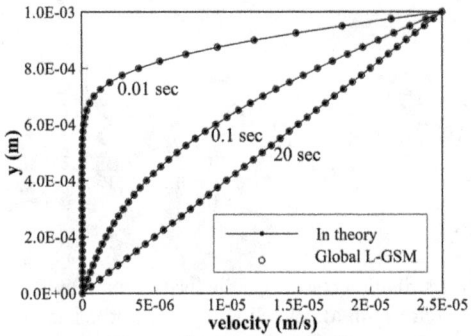

(b) Global L-GSM

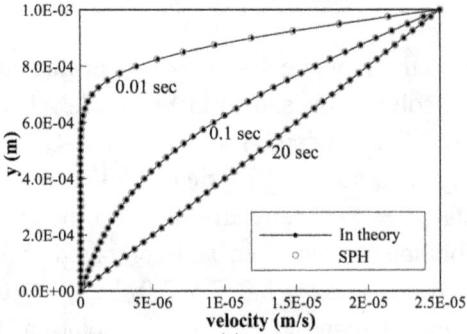

(c) SPH

Figure 7.3. Velocity profiles in the Couette flow along the vertical central line (Reprinted with permission from [5]).

Table 7.2. L_2 error of local L-GSM, global L-GSM, and SPH method for the Couette flow [5].

Δx [m] [$\times 10^{-6}$]	Global L-GSM [$\times 10^{-4}$]	Local L-GSM [$\times 10^{-4}$]	SPH [$\times 10^{-4}$]
5	3.41	3.574	3.441
2.5	1.705	1.785	1.72
2	1.364	1.428	1.376
1.25	0.8525	0.8925	0.86
1	0.682	0.714	0.688
0.5	0.341	0.357	0.344

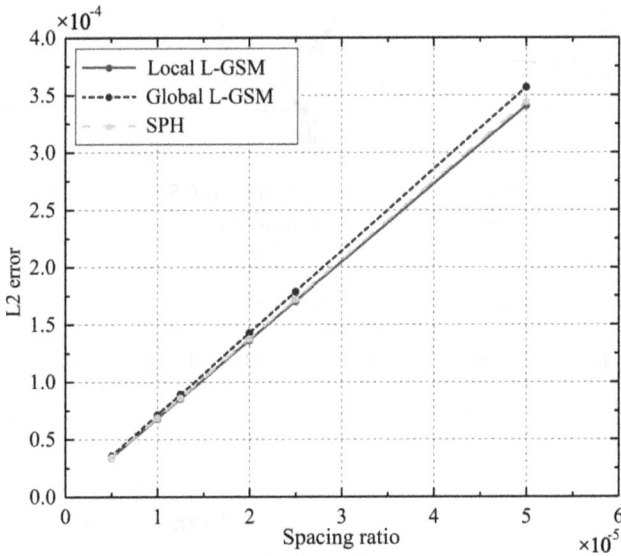

Figure 7.4. Comparison of L_2 error in velocity against particle spacing for the Couette flow (Reprinted with permission from [5]).

7.2.2 Poiseuille flow

For further validation, the L-GSMs are applied to the Poiseuille flow problem. This fluid is initially at rest between two stationary infinite plates. The flow is generated by a constant uniform horizontal force F acting on a fluid.

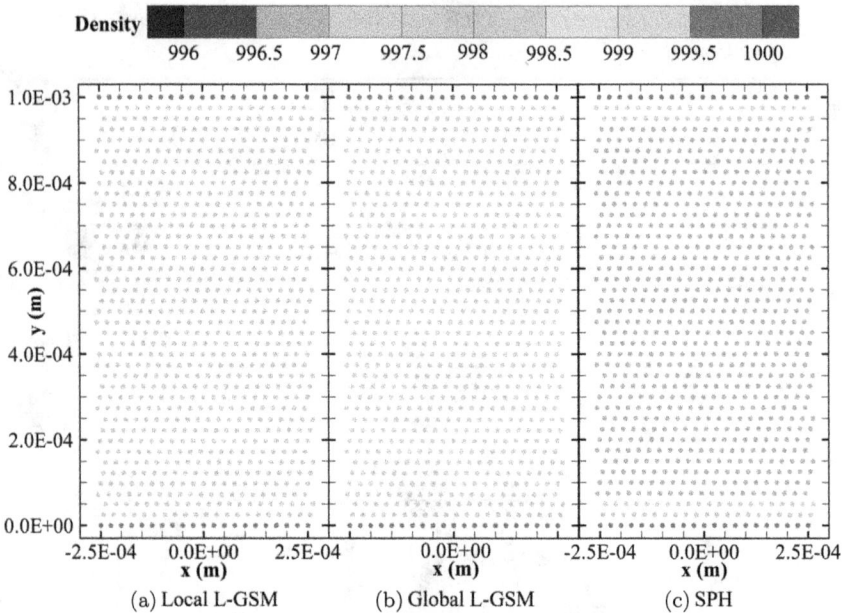

Figure 7.5. Density field of the Couette flow by local L-GSM (a), global L-GSM (b), and SPH (c) at $t = 1.0$ (Reprinted with permission from [6]).

Table 7.3. Parameter settings in Poiseuille flow (mass is for a single particle) [5].

Δx [m]	L [m]	ρ [kg/m^3]	Mass [kg]	v [m^2/s]	F [m/s^2]	c_s [m/s]	Δt[s]
2.5×10^{-5}	10^{-3}	10^3	6.25×10^{-7}	10^{-6}	2.5×10^{-5}	10^{-4}	10^{-4}

The transient velocity field is given in series form by Morris *et al.* [1].

$$\mathbf{v}_x(y,t) = \frac{F}{2v}y(y-L) + \sum_{n=0}^{\infty} \frac{4FL^2}{v\pi^3(2n+1)^3} \sin\left(\frac{\pi y}{L}(2n+1)\right)$$

$$\times \exp\left(-\frac{(2n+1)^2\pi^2 v}{L^2}t\right) \tag{7.8}$$

The setup for numerical simulations is shown in Figure 7.1, and the parameters used are listed in Table 7.3.

The particle distributions and velocity vectors at $t = 20$ for the Poiseuille flow by numerical simulations are shown in Figure 7.6. Good

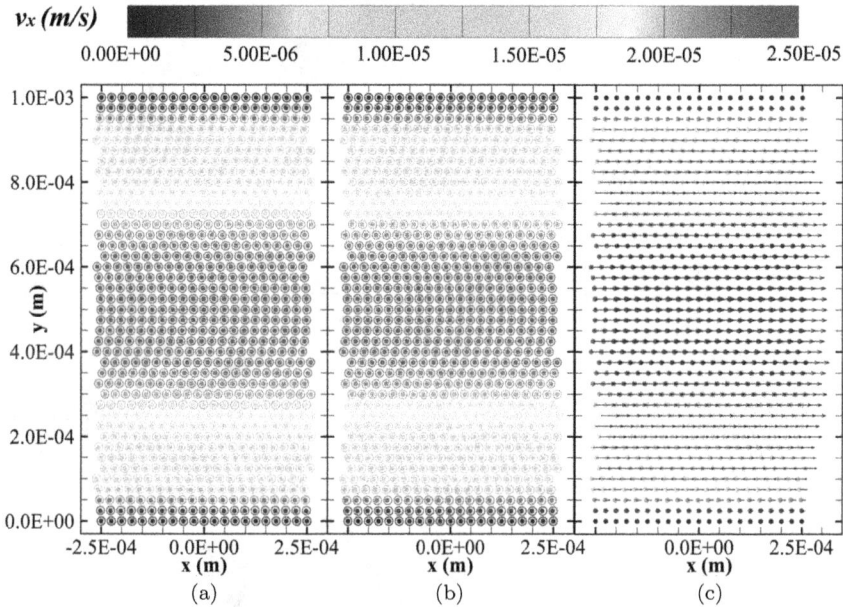

Figure 7.6. Particles distribution comparison in the Poiseuille flow at $t = 1.0$ by local L-GSM (dots in a, b, c), global L-GSM (circles in a), SPH (circles in b), and the velocity quiver by local L-GSM shown in (c) (Reprinted with permission from [5]).

agreement among all three methods is observed. Velocity profiles along the centrally vertical line at different stages obtained using different methods are compared to the theoretical result in Figure 7.7. A close agreement with a maximum discrepancy of 0.5% is found for the steady-state solution.

Simulations are also conducted using different numbers of particles spanning the channel. The relative errors are listed in Table 7.4 and plotted in Figure 7.8. The error profile in Figure 7.8 confirms that all three methods, local L-GSM, global L-GSM, and SPH, are 1st order accurate.

7.2.3 Shear-driven cavity

The L-GSM is further validated using the shear-driven cavity problem as the third example. The fluid flows in a closed square with a moving top sheet at a constant velocity V_{top}, while the other three sides remain stationary. The flow will reach a steady state and eventually form a recirculation pattern.

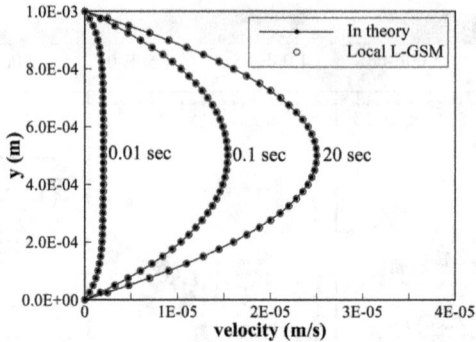

(a) Local L-GSM

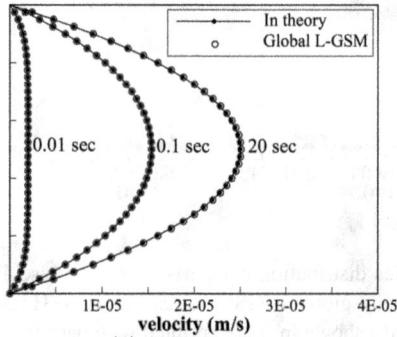

(b) Global L-GSM

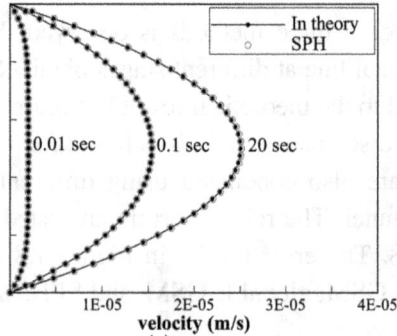

(c) SPH

Figure 7.7. Velocity profile of Poiseuille flow along the vertical central line by local L-GSM/global L-GSM/SPH at $t = 0.01$, 0.1 and 20, respectively (Reprinted with permission from [5]).

Table 7.4. L_2 error of local L-GSM, global L-GSM, and SPH method in solving the Poiseuille flow [5].

Δx [m] $[\times 10^{-6}]$	Global L-GSM $[\times 10^{-4}]$	Local L-GSM $[\times 10^{-4}]$	SPH $[\times 10^{-4}]$
5	5.771	5.860	6.577
2.5	2.885	2.930	3.285
2	2.308	2.344	2.628
1.25	1.443	1.465	1.643
1	1.154	1.172	1.314
0.5	0.577	0.586	0.657

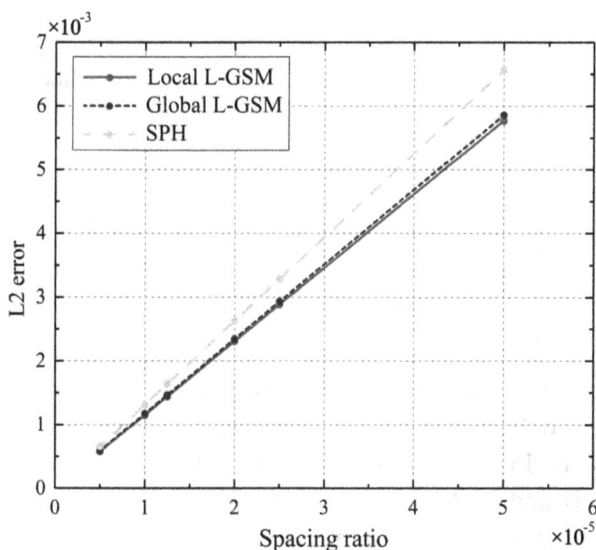

Figure 7.8. Comparison of L_2 solution error against particle spacing for the Poiseuille flow (Reprinted with permission from [5]).

A total of 1600 real particles are arranged in the square domain and 324 boundary particles are placed on the four boundaries, as shown in Figure 7.9. The global L-GSM model based on the governing equations in Eq. (7.1) and EoS (7.2), and the no-slip solid boundary treatment presented in Section 6.8.2, is compared to that of the SPH model and a FDM model [7]. The parameters adopted in simulation are listed in Table 7.5.

Real particle   No-slip boundaryparticle

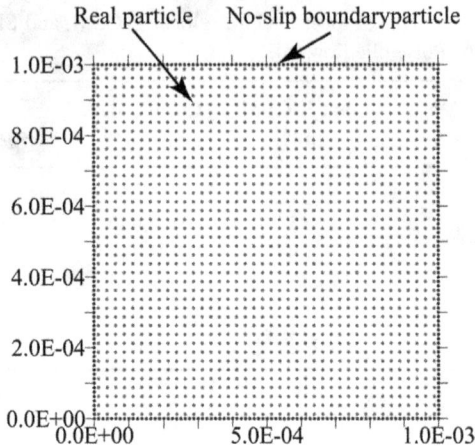

Figure 7.9. Initial particle distribution. The interior real particles (red dots) and the boundary particles (blue dots) are shown (Reprinted with permission from [4]).

Table 7.5. Parameter settings for shear-driven cavity problem [4].

Δx [m]	L [m]	ρ [kg/m^3]	Mass [kg]	ν [m^2/s]	V_{Top} [m/s]	c_s [m/s]	Δt [s]
2.5×10^{-5}	10^{-3}	10^3	6.25×10^{-7}	10^{-6}	10^{-3}	10^{-2}	5×10^{-5}

The velocity quivers at the steady state and the velocity along the centerline obtained from these three numerical approaches is compared in Figure 7.10 and Figure 7.11, respectively. The L-GSM results agree well with the SPH and FDM results. The L-GSM slightly underpredicts the velocity compared to the Eulerian FDM results. This is mainly due to the fact that the velocity on the centerline is taken in FDM, while the values of velocity in the L-GSM are approximated by those at the particles nearest the centerline.

7.2.4 *Dam break*

The L-GSM is used to simulate the flow from a broken dam. The water flow is subjected to gravity, and the upper surface of the water is free of pressure. The results are compared to that of an SPH model [8]. The simulation involves 2500 real particles and 230 boundary particles, distributed

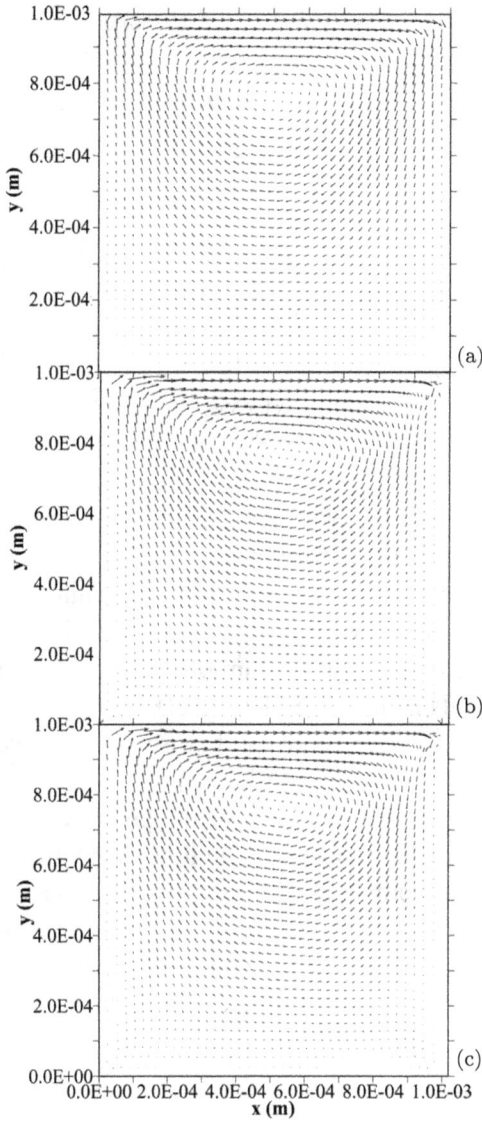

Figure 7.10. Steady-state velocity quiver in the cavity flow using FDM (a), L-GSM (b), and SPH (c). The arrow length represents the magnitude of velocity (Reprinted with permission from [4]).

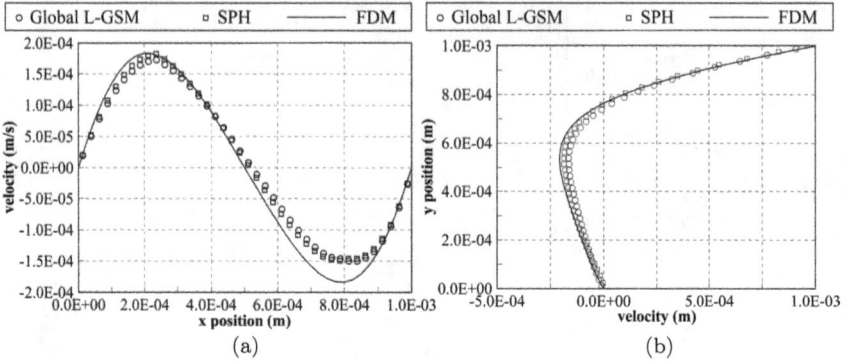

Figure 7.11. Comparison of vertical velocities along the horizontal centerline (a), and horizontal velocities along the vertical centerline (b) (Reprinted with permission from [4]).

in a range of 25 m × 25 m. The same set of initial particles is used in all numerical models for fair comparisons.

The first case sets an initially static water column. It is released at the first time step, and the flow hits a free-slip obstacle placed 50 m away from the left wall of the water column. Figure 7.12 compares the flowing processes simulated by L-GSM and SPH at four time instances, which demonstrates the effectiveness of the free surface treatment presented in Section 6.3.1.

In the second case, the evolution of the profile of the fluid flow over time is studied. The time histories of the flow's height H and surging front Z are computed and compared with experimental data in Figure 7.13 and Table 7.6. Results show the agreement between the L-GSM and experimental results, indicating the model's ability to handle dynamic large movement of fluids with evolving free surface topography.

7.2.5 *Wall impact of breaking dam*

This test examines the behavior of incompressible free surface flow with a relatively high-velocity impacting a wall at distance of 80 m from the left wall of the water column. The initial configuration of the problem is the same as the previous case of dam break.

The velocity field and particle configuration at time instances obtained by L-GSM and SPH are compared in Figure 7.14. The selected time

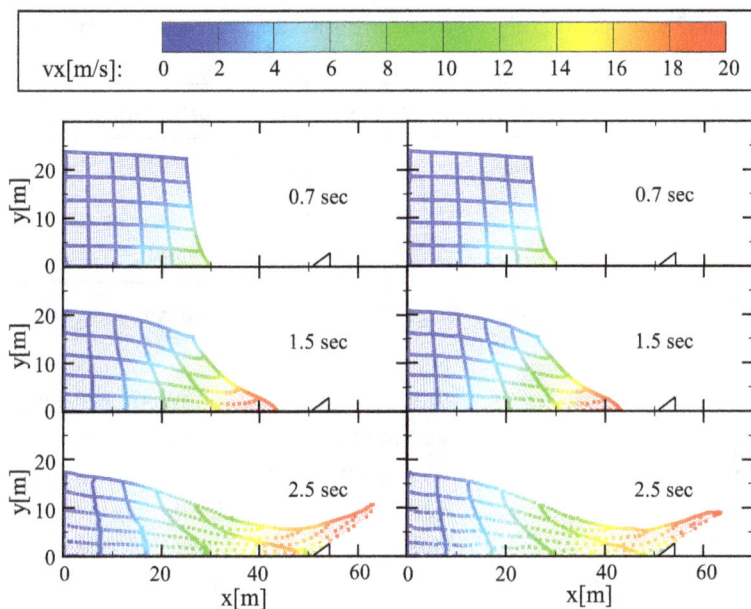

Figure 7.12. Comparison of velocity field and the distribution of 2730 particles at the time instances, after the dam breaks. L-GSM (left) and SPH (right). v_x is the velocity in horizontal x-direction (Reprinted with permission from [3]).

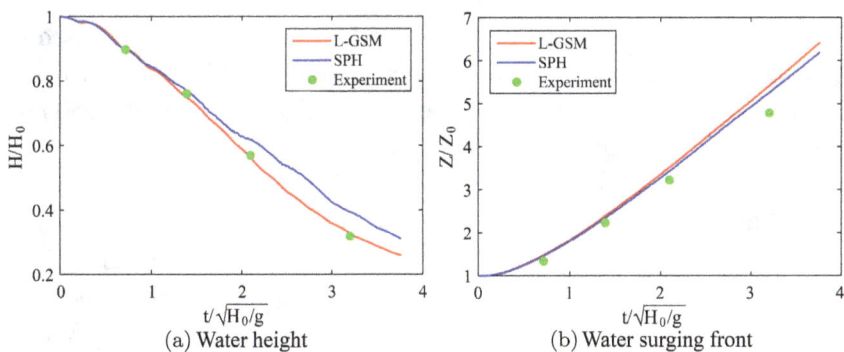

(a) Water height   (b) Water surging front

Figure 7.13. Water height (in unit of initial height 25 m) and water surging front (in unit of initial column length 25 m) evolve with time (in unit of $\sqrt{H_0/g}$), after the dam breaks. Comparison of L-GSM, SPH, and experiment without obstacle (Reprinted with permission from [3]).

Table 7.6. Values of height and surging front of the dam break by experiment, L-GSM, and SPH [3].

Time	HT_{exp}	$HT_{L\text{-}GSM}$	HT_{SPH}	SR_{exp}	$SR_{L\text{-}GSM}$	SR_{SPH}
0.71	0.90	0.90	0.90	1.33	1.48	1.46
1.39	0.76	0.75	0.77	2.25	2.40	2.35
2.10	0.57	0.56	0.61	3.22	3.53	3.44
3.20	0.32	0.33	0.39	4.80	5.43	5.27

Note: HT-Height of water; SR-Surge front of water; exp-Experiment data [9].

instances include the first wall impact at $t = 3.2$, the maximum height of the flow at $t = 6.5$, a stage between the first two wall impacts at $t = 10$, and after the second impact at $t = 15$. As shown in Figure 7.15, L-GSM captures the velocity vortex more accurately. This is due to the more precise free surface treatment in L-GSM. Figure 7.16 compares the time history of the maximum height and maximum surging front in the dam breaking and wall impact processes by L-GSM and SPH, and both solutions show basic agreement.

7.2.6 *Water discharge*

To further validate the L-GSM model for free surface flows, it is used to simulate a water discharge process, together with the SPH for comparison. These numerical models use the same set of initial particles with 1250 real and 152 boundary particles, as shown in Figure 7.17.

Results from the simulations are shown in Figure 7.18, which reveals a good agreement between the two methods for both the velocity field of water in the container and the discharge profile beyond the container. The L-GSM solution shows a more flexible discharge profile beyond the container than SPH, consistent with earlier observations. The velocity vector in Figure 7.19 again confirms the reliability of the L-GSM model in simulating free surface flows. Figure 7.20 compares the maximum water level within the container and the horizontal velocity near the "outlet line" as indicated in Figure 7.17, obtained by L-GSM and SPH. The water level by L-GSM shows good agreement with the SPH result, and the velocity profiles by both numerical methods have similar trends on average.

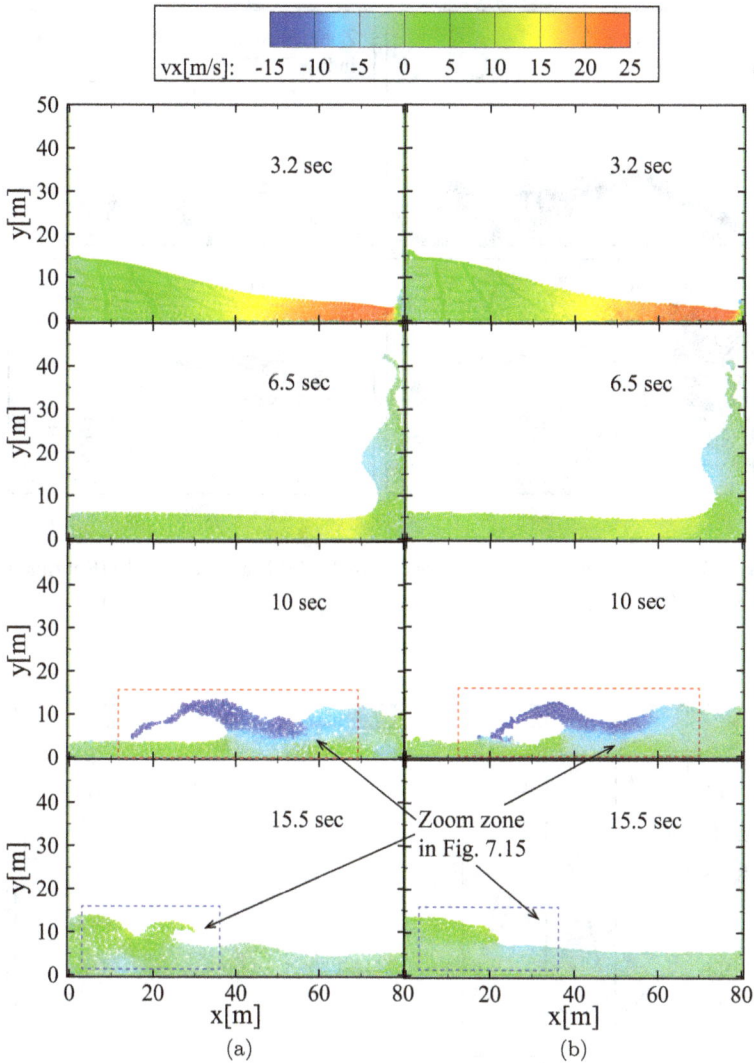

Figure 7.14. Comparison of velocity field and particle distribution at multiple time instances after the dam breaks, obtained by L-GSM (a) and SPH (b). v_x is the velocity in horizontal x-direction (Reprinted with permission from [3]).

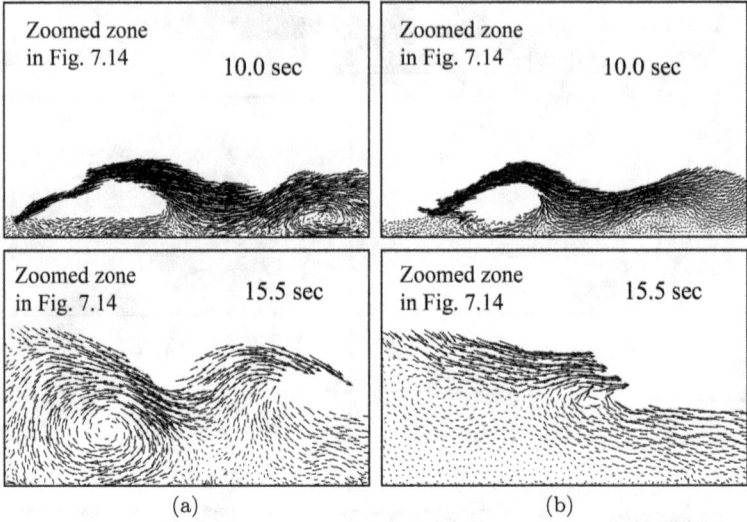

Figure 7.15. Comparison of velocity vector by L-GSM (a) and SPH (b) (Reprinted with permission from [3]).

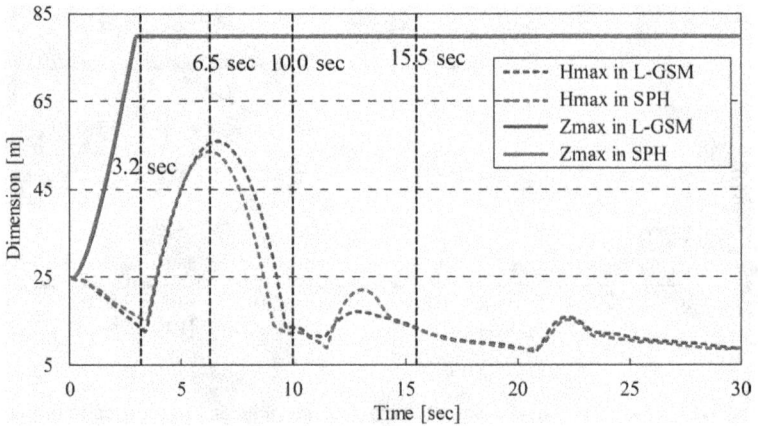

Figure 7.16. History of maximum height and surging front and wall impact processes after the dam breaks. The selected time instances shown in Figure 7.14 are marked by these vertical dashed lines (Reprinted with permission from [3]).

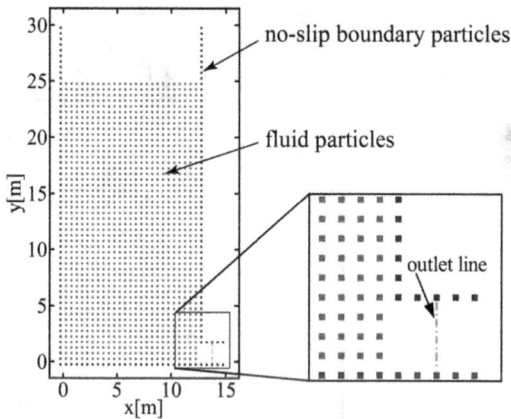

Figure 7.17. Particle configuration for water discharge simulation. Red squares: fluid particles; blue squares: free-slip solid boundary particles simulating the container wall; dashed-dot line: the "outlet line" where the velocity in horizontal direction is investigated and plotted in Figure 7.20 (Reprinted with permission from [3]).

7.2.7 Droplet splash

This study simulates the impact of a sphere droplet on the surface of rest water in an infinitely long channel, using both the L-GSM and SPH. The same set of particles is used for both methods, which consists of 1000 fluid particles representing the rest water in the channel, 61 fluid particles representing the incoming water droplets, and 116 boundary particles representing the free-slip solid wall as illustrated in Figure 7.21. The velocity contour and velocity vector of the water splash process by L-GSM are presented in Figures 7.22 and 7.23, respectively. Both the velocity contour and the velocity vector are found to be smooth and make good sense physically. This demonstrates yet again the reliability of the L-GSM model for simulating free surface flow.

The L-GSM solution to the maximum water heights at two selected positions in the entire process is quantitatively validated by comparing to the corresponding SPH solution, as shown in Figure 7.24. It can be observed that the height histories obtained from both numerical methods exhibit a similar period and oscillation amplitude. The L-GSM results are smoother than those of SPH. At last, the L-GSM simulation results with 25,000 real particles are presented in Figure 7.25.

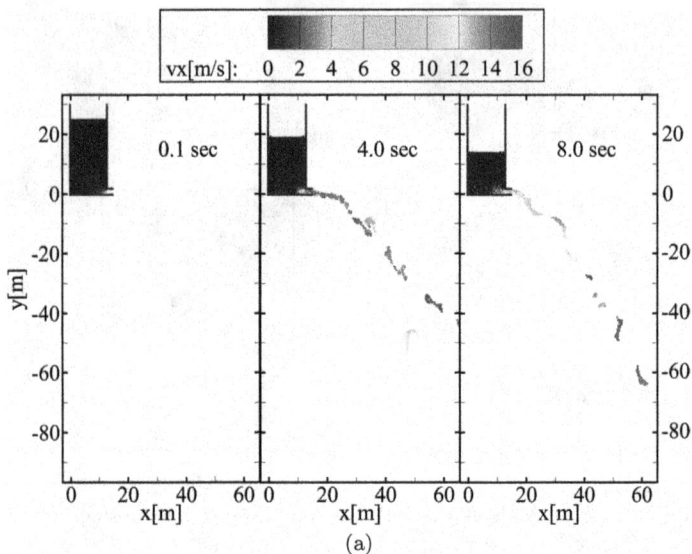

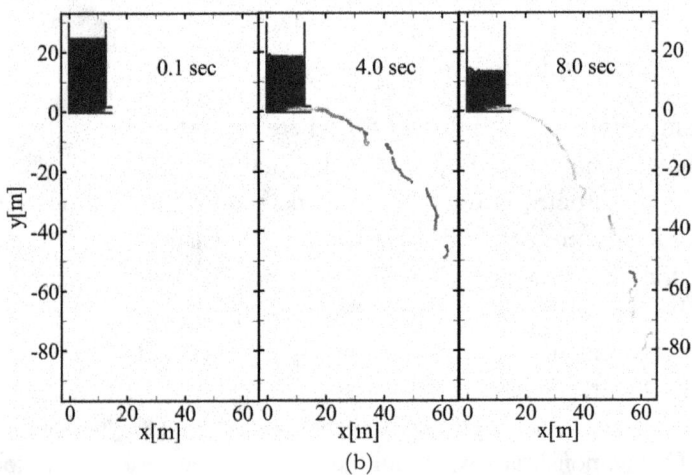

Figure 7.18. Velocity field and particle distribution during the water discharge process, obtained by L-GSM (a) and SPH (b) at the selected time instances. v_x is the horizontal velocity (Reprinted with permission from [3]).

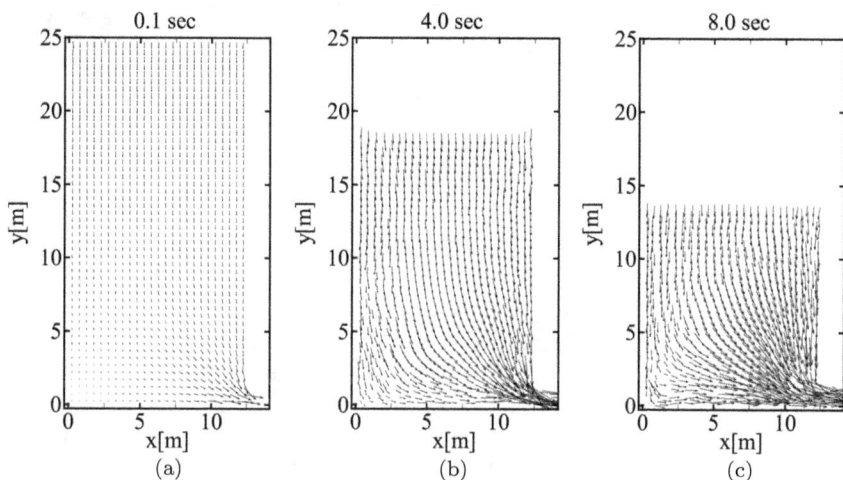

Figure 7.19. Velocity vector of water discharge by L-GSM (Reprinted with permission from [3]).

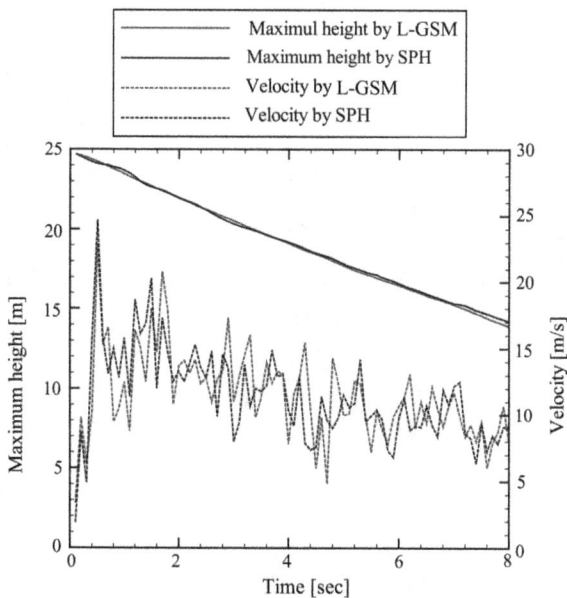

Figure 7.20. Comparison of maximum water level in the container (solid lines), and the horizontal velocity near the "outlet line" indicated in Figure 7.18 (dashed lines). L-GSM (red color). SPH (blue color) (Reprinted with permission from [3]).

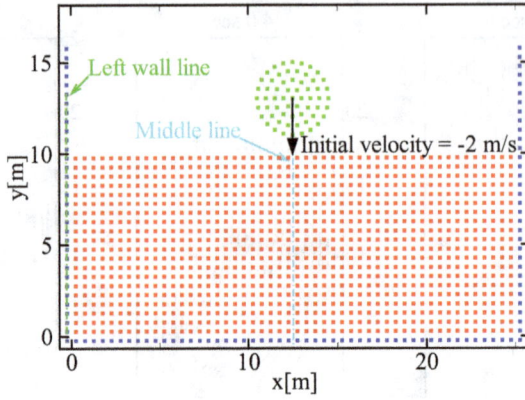

Figure 7.21. Numerical model of water splash. Red: rest water in container. Green: falling droplet with a velocity of 2 m/s. Blue: solid boundary particles (Reprinted with permission from [3]).

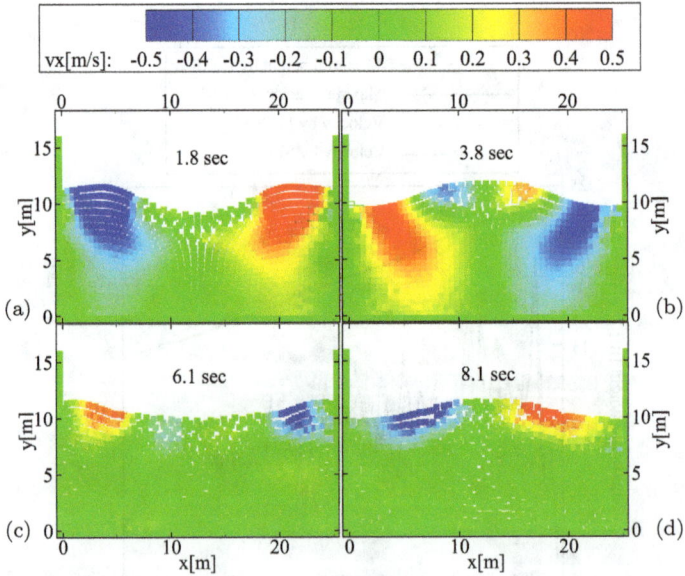

Figure 7.22. Velocity contours of water splash simulated using L-GSM. v_x is the velocity in horizontal direction (Reprinted with permission from [3]).

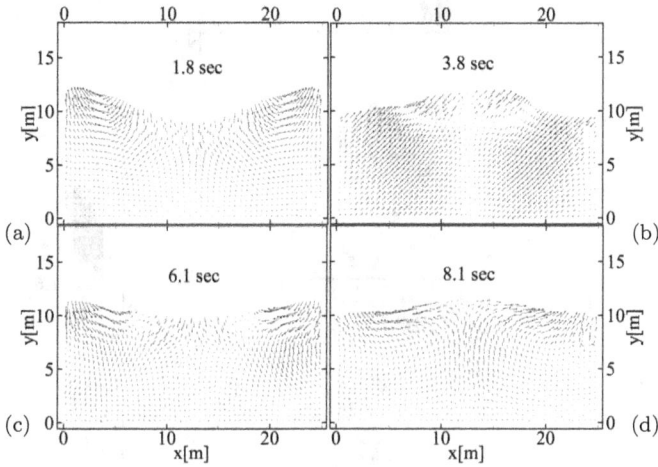

Figure 7.23. Velocity vector of water splash by L-GSM at $t = 1.8$ sec, 3.8 sec, 6.1 sec, and 8.1 sec (Reprinted with permission from [3]).

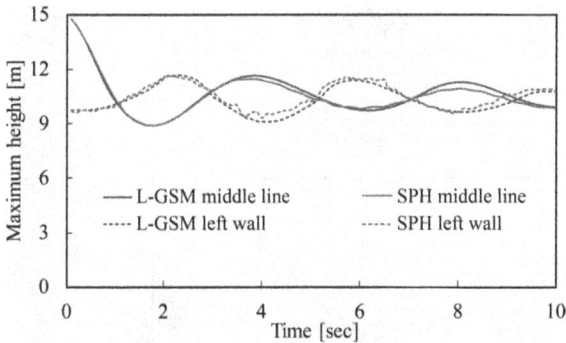

Figure 7.24. Comparison of maximum water height at the positions of middle line (solid curves) and left wall line (dashed curves) by L-GSM (red) and SPH (green). The middle line and left wall line are indicated in Figure 7.21 (Reprinted with permission from [3]).

Based on the applications presented, it may be concluded that both the global L-GSM and local L-GSM are capable of producing reliable numerical solutions compared to existing theoretical and numerical solutions. The non-conservation of momentum in the local L-GSM does not significantly impact the accuracy of the numerical solutions.

The next chapter presents applications of L-GSM to solve granular flows in geotechnical engineering.

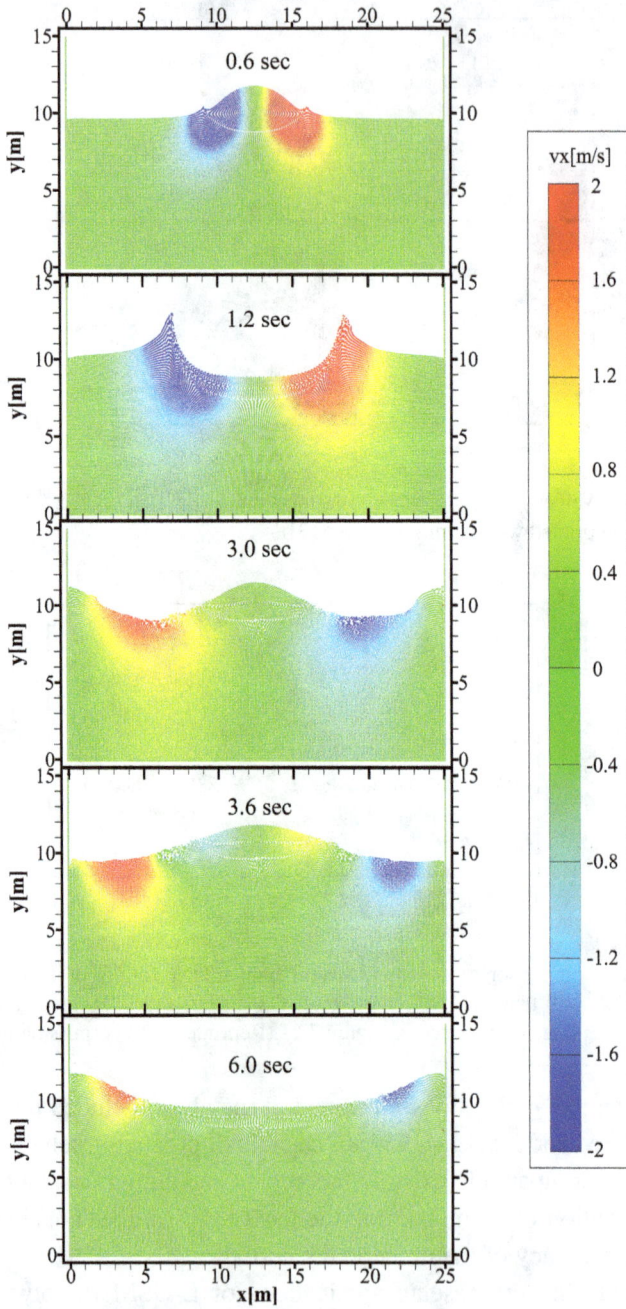

Figure 7.25. L-GSM simulation of splash with 25,000 particles.

References

[1] J. P. Morris, P. J. Fox, and Y. Zhu, Modeling Low Reynolds Number Incompressible Flows Using SPH, *Journal of Computational Physics*, 136, 214–226 (1997).

[2] J. J. Monaghan, Simulating Free Surface Flows with SPH, *Journal of Computational Physics*, 110(2), 399–406 (1994, February).

[3] Z. Mao, G. R. Liu, X. Dong, and T. Lin, "A conservative and consistent Lagrangian gradient smoothing method for simulating free surface flows in hydrodynamics," *Computational Particle Mechanics*, 6, 781–801 (2019). doi: 10.1007/s40571-019-00262-z.

[4] Z. Mao and G. R. Liu, A Lagrangian gradient smoothing method for solid-flow problems using simplicial mesh, *International Journal for Numerical Methods in Engineering*, 113(5), 858–890 (2018, February). doi: 10.1002/nme.5639.

[5] Z. Mao, G. R. Liu, and Y. Huang, A local Lagrangian gradient smoothing method for fluids and fluid-like solids: A novel particle-like method, *Engineering Analysis with Boundary Elements*, 107, 96–114 (2019, October). doi: 10.1016/j.enganabound.2019.07.003.

[6] Z. Mao, A Novel Lagrangian gradient smoothing method for fluids and flowing solids, *University of Cincinnati*, (2019). Accessed: January 04, 2021. Available: http://rave.ohiolink.edu/etdc/view?acc_num=ucin1553252214052311.

[7] K. Poochinapan, -H Hsu, M. Langthjem, and M. Mei, Numerical Implementations for 2D Lid-Driven Cavity Flow in Stream Function Formulation, *International Scholarly Research Network ISRN Applied Mathematics*, 2012, 17 (2012). doi: 10.5402/2012/871538.

[8] G. R. Liu and M. B. Liu, *Smoothed Particle Hydrodynamics: A Meshfree Particle Method*. Singapore: World Scientific Publishing Co (2003).

[9] J. C. Martin and W. J. Noyce, *Philosophical Transactions of the Royal Society*, 244, 312 (1952).

Chapter 8

L-GSM for Granular Flows in Geotechnical Engineering

Contents

This chapter focuses on applications of global and local L-GSMs to challenging problems in geotechnical engineering. We aim to simulate flowing geological materials, such as cohesive soil, which undergo large deformation. The major challenge in dealing with this type of problem is that cohesive soil is subjected to tensile stress, resulting in tensile instability in numerical simulations. The SPH has been found to suffer from such instability and hence limited its applications. This study will show the excellent ability of the L-GSMs in overcoming such instability, which are ideal methods to solve this type of challenging problem. This chapter is organized as follows:

- Section 8.1 introduces the discretized form of governing and constitutive equations for geological materials.

- Section 8.2 presents simulation results using both global L-GSM and local L-GSM for a number of problems of flowing geological materials that undergo large deformation, including

  o non-cohesive soil column collapse
  o cohesive soil column collapse
  o Earthquake-induced landslides

- Section 8.3 analyzes the computational efficiency of global L-GSM and local L-GSM, in terms of accuracy, adaptability, stability, and efficiency.
- Finally, Section 8.4 provides concluding remarks.

8.1 Constitutive and governing equations

Geological materials are essentially elastoplastic. Materials considered in this study are all assumed to be elastic perfectly plastic. The failure behavior is also complicated. In this study, a constitutive model associated with the Drucker–Prager yield criterion is used, which has also been employed in SPH simulations for soil problems [1]. Here, we provide the final formulation of constitutive equations. Readers may refer to [1] for the detailed derivation.

The stress–strain relationship is expressed as

$$\frac{D\sigma_i^{\alpha\beta}}{Dt} = \sigma_i^{\alpha\gamma}\dot{\omega}_i^{\beta\gamma} + \sigma_i^{\gamma\beta}\dot{\omega}_i^{\alpha\gamma} + K_i\dot{\varepsilon}_i^{\gamma\gamma}\delta^{\alpha\beta} + 2G_i\dot{e}_i^{\alpha\beta}$$
$$-\dot{\lambda}_i\left[3K_i\alpha_{\psi i}\delta^{\alpha\beta} + \left(\frac{G}{\sqrt{J_2}}\right)_i\tau_i^{\alpha\beta}\right] \tag{8.1}$$

in which

$$\dot{\lambda}_i = \frac{3\alpha_{\phi i}K_i\dot{\varepsilon}_i^{\gamma\gamma} + (G/\sqrt{J_2})_i\tau_i^{\alpha\beta}\dot{\varepsilon}_i^{\alpha\beta}}{9K_i\alpha_{\phi i}\alpha_{\psi i} + G_i} \tag{8.2}$$

where α, β, and γ are indexes for the Cartesian components x, y, and z with Einstein summation implied; $\delta^{\alpha\beta}$ is the Kronecker delta; K and G are the elastic bulk and shear moduli of the material, respectively; $\tau^{\alpha\beta}$ is the deviatoric stress; and $\dot{e}^{\alpha\beta}$ is the deviatoric shear strain rate given by

$$\dot{e}^{\alpha\beta} = \dot{\varepsilon}^{\alpha\beta} - \frac{1}{3}\dot{\varepsilon}^{\gamma\gamma}\delta^{\alpha\beta} \tag{8.3}$$

$$\dot{\varepsilon}^{\alpha\beta} = \frac{1}{2}\left(\frac{\partial v^\alpha}{\partial x^\beta} + \frac{\partial v^\beta}{\partial x^\alpha}\right) \tag{8.4}$$

$\dot{\omega}^{\alpha\beta}$ is the spin rate tensor:

$$\dot{\omega}^{\alpha\beta} = \frac{1}{2}\left(\frac{\partial v^\alpha}{\partial x^\beta} - \frac{\partial v^\beta}{\partial x^\alpha}\right) \tag{8.5}$$

The last term in Eq. (8.1) is determined according to the following yield criterion:

$$\begin{cases} \dot{\lambda} = 0, & f < 0 \\ \dot{\lambda} > 0, & f = 0 \end{cases} \tag{8.6}$$

with the yield function f being

$$f = \alpha_\phi I_1 + \sqrt{J_2} - k_c \tag{8.7}$$

where I_1 is the first invariant of the stress tensor and J_2 is the second invariant of the deviatoric stress tensor, α_ϕ and k_c are Drucker–Prager constants that are determined using the experimentally measurable Coulomb material constant c (cohesion) and ϕ (internal friction angle). For 2D plane strain cases,

$$\alpha_\phi = \frac{\tan\phi}{\sqrt{9+12\tan^2\phi}} \quad k_c = \frac{3c}{\sqrt{9+12\tan^2\phi}} \tag{8.8}$$

The invariants of stress tensors are computed using

$$I_1 = \frac{3}{2}(\sigma_{xx} + \sigma_{yy}) - 3\alpha\sqrt{J_2} \quad J_2 = \left[\left(\frac{\sigma_{xx}+\sigma_{yy}}{2}\right)^2 + \sigma_{xy}^2\right]\Big/(1-3\alpha^2) \tag{8.9}$$

For 3D cases,

$$\alpha_\phi = \frac{2\sin\phi}{\sqrt{3}(3-\sin\phi)} \quad k_c = \frac{6c\cos\phi}{\sqrt{3}(3-\sin\phi)} \tag{8.10}$$

$$I_1 = \sigma_{xx} + \sigma_{yy} + \sigma_{zz}$$

$$J_2 = \frac{1}{6}[(\sigma_{xx} - \sigma_{yy})^2 + (\sigma_{yy} - \sigma_{zz})^2 + (\sigma_{zz} - \sigma_{xx})^2] + \sigma_{xy}^2 + \sigma_{yz}^2 + \sigma_{zx}^2$$

$$(8.11)$$

The dilatancy factor α_ψ in Eq. (8.1) is related to the dilatancy angle Ψ in a fashion similar to α_ϕ relating to friction angle ϕ.

Equation (8.1) contains a total of six material parameters: density (ρ), cohesion (c), internal friction angle (ϕ), dilatancy angle (ψ), elastic modulus (E), and Poisson's ratio (v). The values of these parameters are listed in tables given in the validation subsections for the materials used in the simulation.

To ensure the material behavior is consistent with this elastic perfectly plastic constitutive model, two return mapping algorithms, named "tension cracking treatment" and "stress scaling back" [1], which can be dated back to the work in FEM simulation [2], are also used in our L-GSM models.

The full set of governing equations associated with the constitutive model is given as follows in discretized form:

$$\begin{cases} \dfrac{D\rho_i}{Dt} = -\dfrac{\rho_i}{V_i} \sum_{k=1}^{n_i} \left(\dfrac{v_{jk}^\alpha + v_i^\alpha}{2} \right) (\Delta S_\alpha)_{ij_k} \\[3mm] \dfrac{Dv_i^\alpha}{Dt} = \dfrac{1}{\rho_i V_i} \sum_{k=1}^{n_k} \left(\dfrac{\sigma_{jk}^{\alpha\beta} + \sigma_i^{\alpha\beta}}{2} - \Pi_{ij_k}\delta^{\alpha\beta} \right)(\Delta S_\alpha)_{ij_k} + F^\alpha \\[3mm] \dfrac{Dd^\alpha}{Dt} = v^\alpha \\[3mm] \dfrac{D\sigma_i^{\alpha\beta}}{Dt} = \sigma_i^{\alpha\gamma}\dot{\omega}_i^{\beta\gamma} + \sigma_i^{\gamma\beta}\dot{\omega}_i^{\alpha\gamma} + K_i\dot{\varepsilon}_i^{\gamma\gamma}\delta^{\alpha\beta} + 2G_i\dot{e}_i^{\alpha\beta} \\[2mm] \qquad\quad - \dot{\lambda}_i\left[3K_i\alpha_{\psi i}\delta^{\alpha\beta} + \left(\dfrac{G}{\sqrt{J_2}}\right)_i \tau_i^{\alpha\beta} \right] \end{cases}$$

$$(8.12)$$

with

$$\dot{\omega}^{\alpha\beta} = \frac{1}{2}\left(\frac{\partial v^\alpha}{\partial x^\beta} - \frac{\partial v^\beta}{\partial x^\alpha} \right)$$

$$= \frac{1}{2}\left[\frac{1}{V_i}\sum_{k=1}^{n_i} \frac{v_{jk}^\alpha + v_i^\alpha}{2}(\Delta S_\beta)_{ij_k} - \frac{1}{V_i}\sum_{k=1}^{n_i} \frac{v_{jk}^\beta + v_i^\beta}{2}(\Delta S_\alpha)_{ij_k} \right] \quad (8.13)$$

$$\dot{\omega}^{\alpha\beta} = \frac{1}{2}\left(\frac{\partial v^\alpha}{\partial x^\beta} - \frac{\partial v^\beta}{\partial x^\alpha}\right)$$

$$= \frac{1}{2}\left[\frac{1}{V_i}\sum_{k=1}^{n_i}\frac{v_{j_k}^\alpha + v_i^\alpha}{2}(\Delta S_\beta)_{ij_k} + \frac{1}{V_i}\sum_{k=1}^{n_i}\frac{v_{j_k}^\beta + v_i^\beta}{2}(\Delta S_\alpha)_{ij_k}\right] \quad (8.14)$$

8.2 Numerical examples

8.2.1 *Soil column collapse*

The soil column collapse under gravity is simulated using three numerical methods: the global L-GSM, local L-GSM, and SPH. The major challenge for simulating this type of geological materials is the possible presence of numerical instability or numerical artifacts. Such artifacts show particularity for soil with cohesivity and hence are capable of taking tensile stress. To reveal and examine this issue clearly, we will conduct simulations for both non-cohesive soil and cohesive soil.

8.2.1.1 *Non-cohesive soil*

2D cases

Consider first non-cohesive soil material $(c = 0)$. The simulation is conducted on a 2D rectangular area, with initial dimensions of 0.2 m length and 0.1 m height. The particle model consists of a total of 3200 real particles with an initial spacing of 0.0025 m. The soil properties used in the simulation are provided in Table 8.1.

The collapsing process of the soil column is shown in Figure 8.1 simulated using global L-GSM and SPH methods. The free surface profile and failure line of the collapsed soil column are compared with the experimental results of Bui *et al.* [1].

Table 8.1.   Soil properties in non-cohesive soil's column [3].

Δx [m]	m [kg]	ρ [kg/m^3]	K [MPa]	v [unit]	c [Pa]	θ [°]	ψ [°]
2.5×10^{-3}	0.01656	2650	0.7	0.3	0	19.8	0

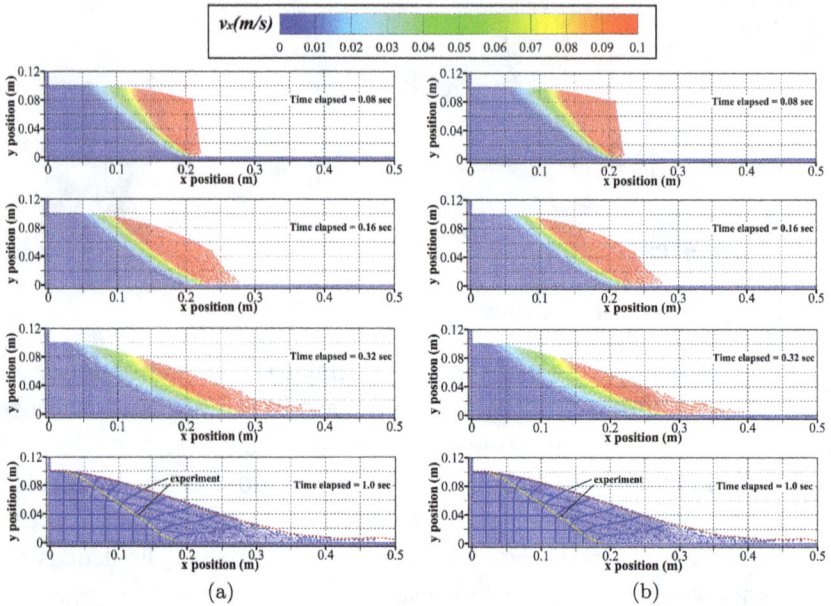

Figure 8.1. Failure process of non-cohesive soil by global L-GSM (a) and SPH (b). The solid and dashed gray curves are the free surface profile and failure line of the soil column, respectively, in the experiment work [1] (Reprinted with permission from [3]).

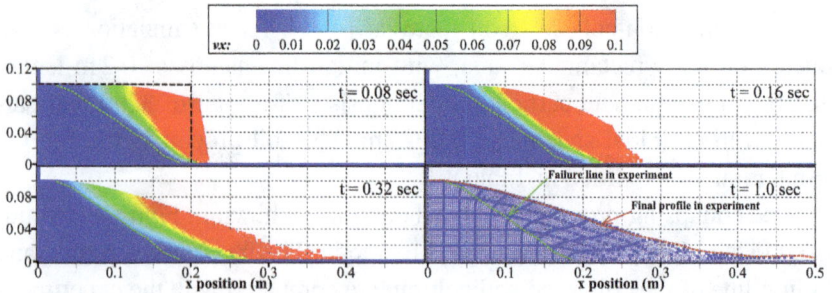

Figure 8.2. Flow process of non-cohesive soil by local L-GSM. v_x is the horizontal velocity (m/s). The dashed curves represent the final profile and the failure line (Reprinted with permission from [4]).

The more efficient local L-GSM is also utilized to simulate the same soil column collapse, and the results are presented in Figure 8.2. The distributions of particles for global L-GSM, local L-GSM, and SPH are compared in Figure 8.3. The results obtained using both the global L-GSM and

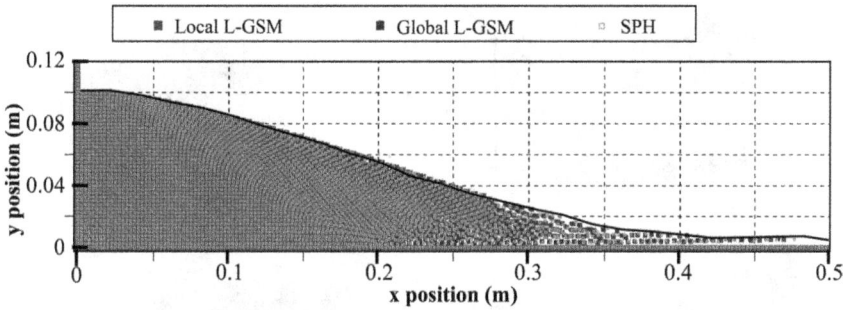

Figure 8.3.   Final profile comparison of non-cohesive soil column collapse obtained by local L-GSM, global L-GSM, and SPH with the experimental result (black curve). The profiles from these three models are very close and almost indistinguishable (Reprinted with permission from [4]).

local L-GSM agree well with those of the SPH method and experimental data.

3D cases

The collapse simulation of non-cohesive soil has also been carried out using the 3D local L-GSM and compared to 3D SPH model using the same set of soil particles. This study was carried out recently and documented in [5].

The soil properties used are listed in Table 8.1. The 3D particle model is shown in Figure 8.4. The study uses a single layer of boundary particles to simulate the solid boundary for local L-GSM, and three layers for SPH. The bottom face is set as "no-slip". The initial size of the soil column is 0.2 m × 0.1 m × 0.1 m, in which the soil particles are spaced 0.01m apart. A total of 2000 soil particles are used for the simulation.

The collapsing process simulated using local L-GSM and SPH is presented in Figure 8.5. The results show that local L-GSM is capable of producing correct flow patterns at different time instances, similar to that of the SPH model.

8.2.1.2   *Cohesive soil*

To investigate the numerical stability of methods for geological materials, cohesive soil needs to be used. This is because the soil cohesivity can

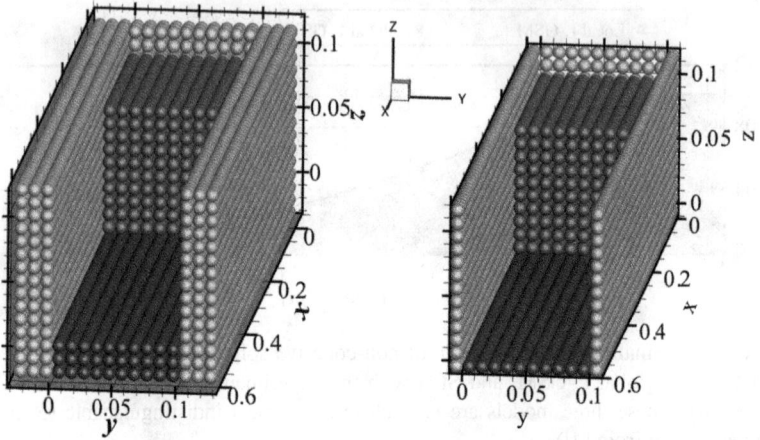

Figure 8.4. Numerical model of 3D SPH (left) and 3D L-GSM (right) (Reprinted with permission from [5]).

trigger numerical instability and lead to spurious non-physical behavior, if the numerical method is not well formulated to handle it.

In this study, we consider cohesive soil and subject it to natural events, such as landslides, which create tension stress in the soil body. The behavior and failure process of the soil are then simulated, using both local L-GSM and global L-GSM, together with SPH. The work was first reported in [4, 6]. The cohesivity of soil is set at $c = 5$ kPa and the internal friction angle is set at $40°$, and the entire set of material property is listed in Table 8.2.

Because SPH is found to have tensile instability for materials with strength, special numerical remedy, such as adding artificial stress into the momentum equation, is needed to suppress the instability [7]. Therefore, four numerical approaches are used for simulating the cohesive soil collapse process:

- SPH without instability remedy,
- SPH with instability remedy,
- global L-GSM without instability remedy,
- local L-GSM without instability remedy.

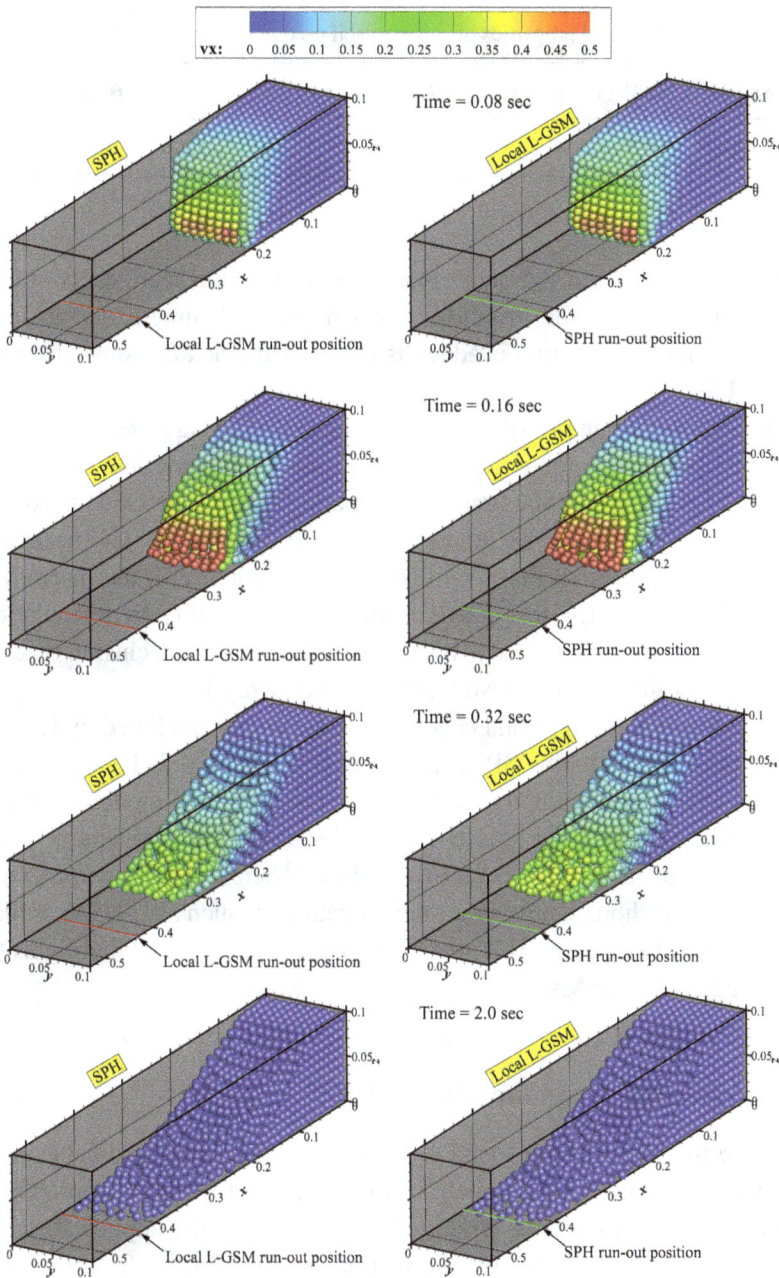

Figure 8.5. Collapsing process of non-cohesive soil column simulated using 3D SPH (left) and 3D L-GSM (right) (Reprinted with permission from [5]).

Table 8.2. Soil properties of cohesive soil used in numerical models [4].

Δx [m]	m [kg]	ρ [kg/m^3]	K [MPa]	v [unit]	c [Pa]	$\theta[^\circ]$	$\psi[^\circ]$
5×10^{-2}	6.625	2650	0.7	0.3	5000	40	0

The results in Figure 8.6 show that SPH without artificial stress suffers from unphysical tensile instability, known also as "numerical cracking". Only when artificial stress is added is unphysical cracking somewhat suppressed.

In contrast, when global L-GSM and local L-GSM are used to simulate the same problem without any special treatment, no numerical cracking is found. This is because the gradient smoothing used in the GSM is consistent approximation for the gradient of the field variables. The necessary physics in the original PDEs will be captured properly in the discretized model. Therefore, L-GSM is free from tensile instability by formulation. No extra treatment is needed. This finding supports the conclusion made in Section 6.11 through a von Neumann stability analysis.

It is found that the frontal edge of the final profile predicted by L-GSM is 3% longer than that by SPH, as presented in Figure 8.7. This is because the artificial stresses added in SPH model made the SPH model stiffer and less fluidic, resulting in a shorter run-out distance. This is the side effect of introducing an artificial stress term into SPH. Therefore, a properly formulated model without using any artificial treatments, such as L-GSM, is ideal for geological materials. This study on cohesive soil material highlights the advantages of L-GSMs.

8.2.2 *Earthquake-induced landslide*

Because the L-GSMs are found to be ideal for simulating the behavior of challenging geological materials, we now apply it to real-life problems. This section examines the performance of the L-GSM applied to large-scale geomechanics problems. We will present simulations for four severe landslides triggered by the 2008 Ms 8.0 Wenchuan earthquake in China. Much of the materials of the section are from the original work [6].

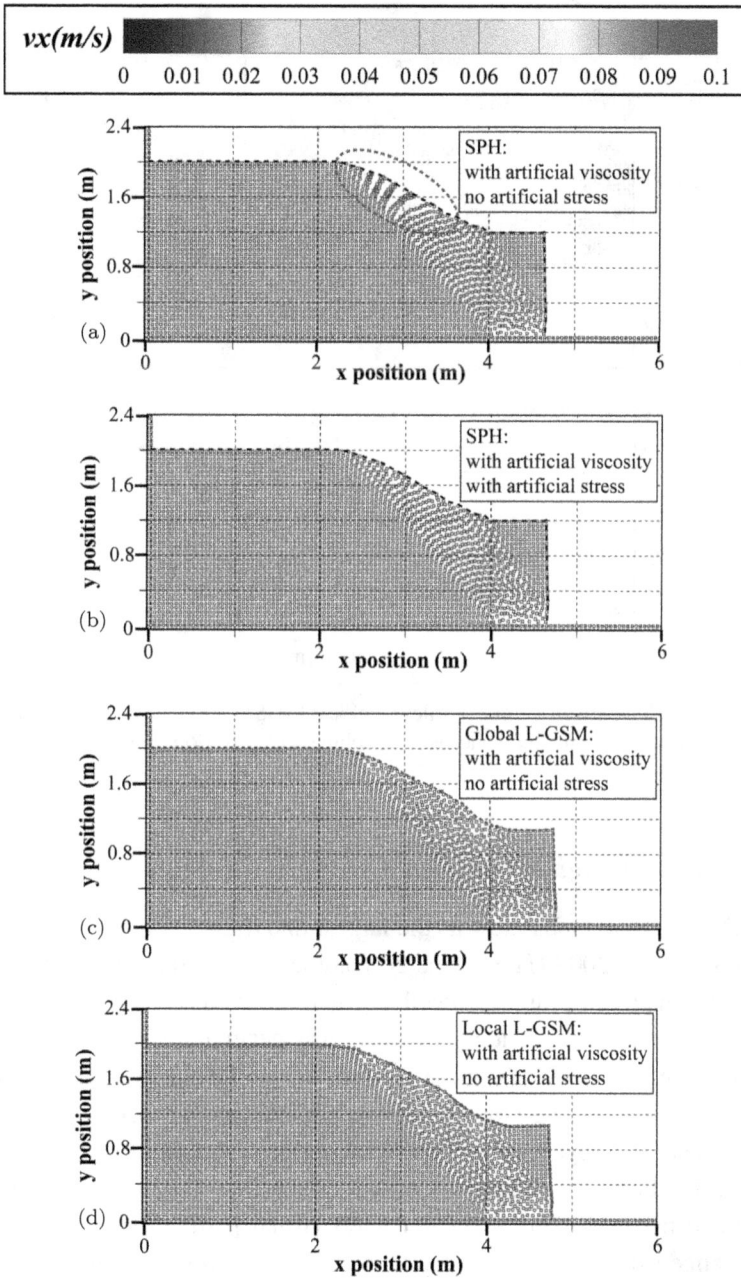

Figure 8.6. Tensile instability behavior in soil simulated using SPH (a-b). Global L-GSM (c) and local L-GSM (d) are free from such instability (Reprinted with permission from [4]).

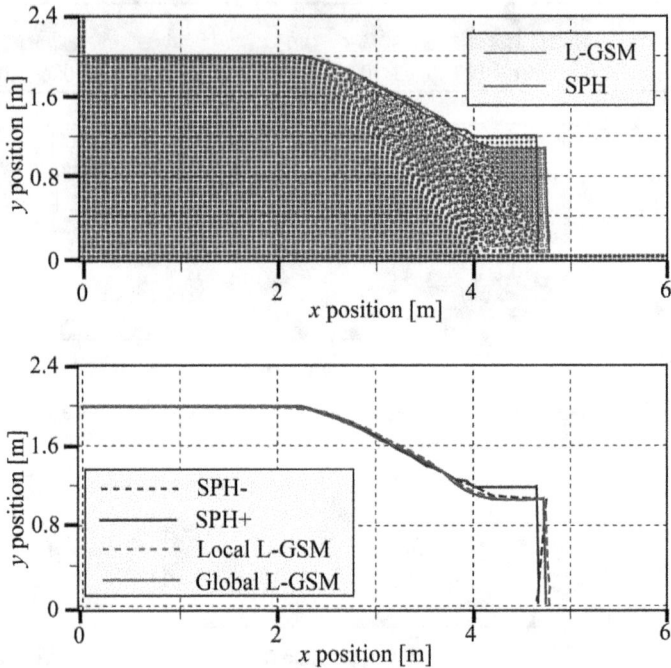

Figure 8.7. Comparison of final profiles by global L-GSM, local L-GSM, and SPH (with artificial stress added) for the cohesive soil column collapse (Reprinted with permission from [4]).

8.2.2.1 *Daguangbao landslide*

The Daguangbao landslide was the largest one triggered by the earthquake in Wenchuan in 2008. The estimated volume of the sliding body was about 800 million m^3. It was simulated using L-GSM. The initial distribution of soil particles and boundary particles was generated based on real data, which were recorded before the earthquake via GPS topography and the onsite surveyed sliding slope as shown in Figure 8.8. A total of 3566 soil particles are employed. At the initial stage, these particles were distributed uniformly above the sliding curve with a spacing of $d = 16$ m. The sliding slope bed was represented by 1380 virtual particles, which horizontally ranged from 0 to 5000 m with a spacing of $d/3$ along the boundary as shown in Figure 8.9. The use of smaller spacing for boundary particles is to better impose the solid boundary condition, by providing a smoother

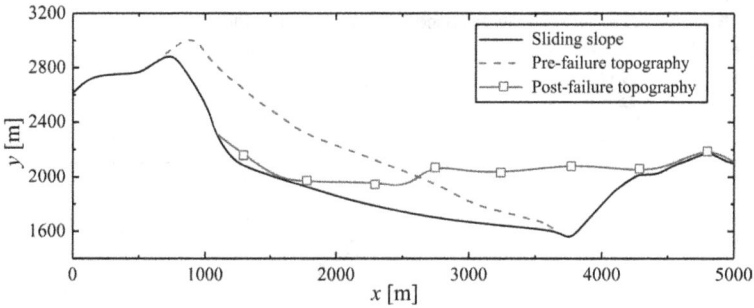

Figure 8.8. Record of pre-failure and post-failure topographies in Daguangbao landslide site (Reprinted with permission from [6]).

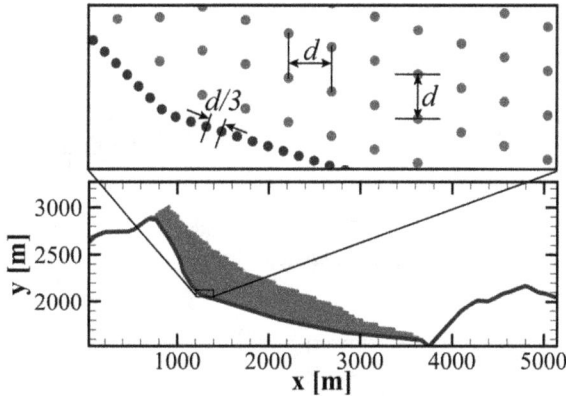

Figure 8.9. Initial particle configuration in L-GSM model for the Daguangbao landslide. The soil particles (in red) have a spacing of $d = 16$ m while the boundary particles (in blue) have a spacing of $d/3$ along the boundary of the slope bed (Reprinted with permission from [9]).

and more consistent repulsive force from the bed. The parameters used in the simulation for the Daguangbao landslide are listed in Table 8.3.

The velocity and pressure fields obtained from L-GSM at four time instances are presented in Figure 8.10 and Figure 8.11, respectively. The post-failure topographies of the landslide obtained using L-GSM and SPH are plotted in Figure 8.12, together with the post-earthquake GPS satellite record. The results show that the boundary treatments on free surface and solid boundary work very well for generating smooth local variable fields near the boundary. The L-GSM result agrees well with that of SPH, and all

Table 8.3. Soil parameters used in Daguangbao landslide simulations [8].

Δx [m]	ρ [kg/m³]	E [GPa]	ν [unit]	c [MPa]	ϕ[°]	ψ[°]
16	2500	1.86	0.2	1.276	10.8	0

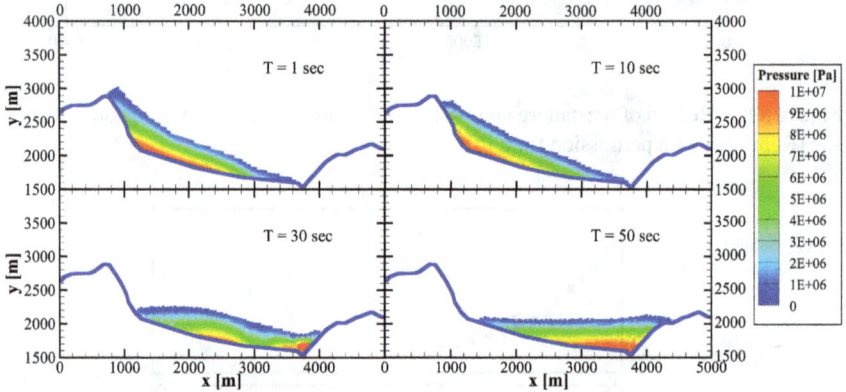

Figure 8.10. Flow process of the Daguangbao landslide simulated by L-GSM model. v_x represents the horizontal velocity (Reprinted with permission from [6])

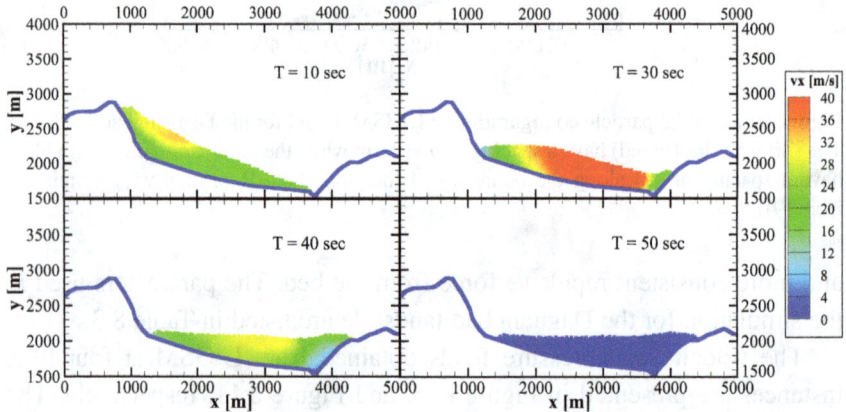

Figure 8.11. Pressure contour of the Daguangbao landslide at four time instances by L-GSM model (Reprinted with permission from [6])

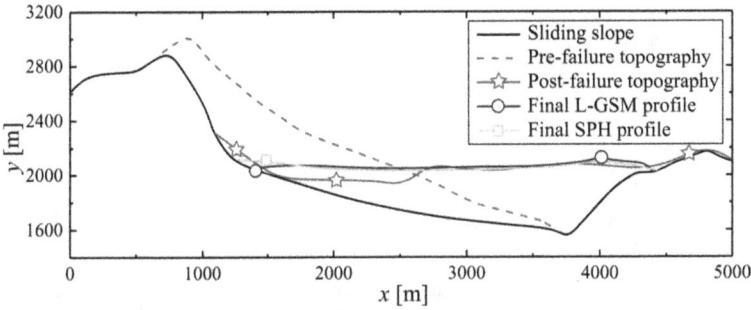

Figure 8.12. Comparison for the final profiles of the Daguangbao landslide simulated using L-GSM and SPH, together with the onsite topography record (Reprinted with permission from [6])

have a general agreement with the recorded data. The difference between the numerical results and the recorded data is attributed to the limitation of 2D for a 3D problem.

8.2.2.2 *Tangjiashan landslide*

The Tangjiashan landslide had a height difference of 650 m between the landslide toe and the main back scar, and a horizontal dimension of 900 m [10]. It was simulated using L-GSM. A total of 3986 soil particles with a spacing of $d = 4$ m and 1622 boundary particles with a spacing of $d/3$ were used. These particles are initially distributed based on the actual recorded sliding slope and pre-failure topography shown in Figure 8.13. The soil parameters used in the simulations are listed in Table 8.4 [11].

A comparison of post-failure topographies by L-GSM, SPH, and the recorded data is shown in Figure 8.13, and the simulated run-out process by L-GSM is presented in Figure 8.14. The L-GSM solution agrees well with both SPH result and the recorded data.

8.2.2.3 *Wangjiayan landslide*

The Wangjiayan landslide in Beichuan County was a high-speed flow-like landslide that occurred during the 2008 Wenchuan earthquake. It resulted in one of the most severe landslide disasters. The sliding body had a volume of 4.8 million m^3, and the landslide occurred on an anti-dip slope.

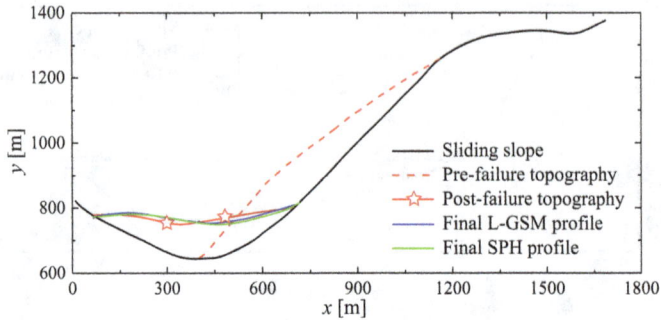

Figure 8.13. Comparison for the final profiles of the Tangjiashan landslide simulated using L-GSM and SPH, together with the onsite topography record (Reprinted with permission from [6])

Table 8.4. Soil parameters used in the simulation for the Tangjiashan landslide simulation [11].

Δx [m]	ρ [kg/m^3]	K [GPa]	G [GPa]	c [kPa]	ϕ [°]	ψ [°]
4	2000	1.3	0.8	30	35	0

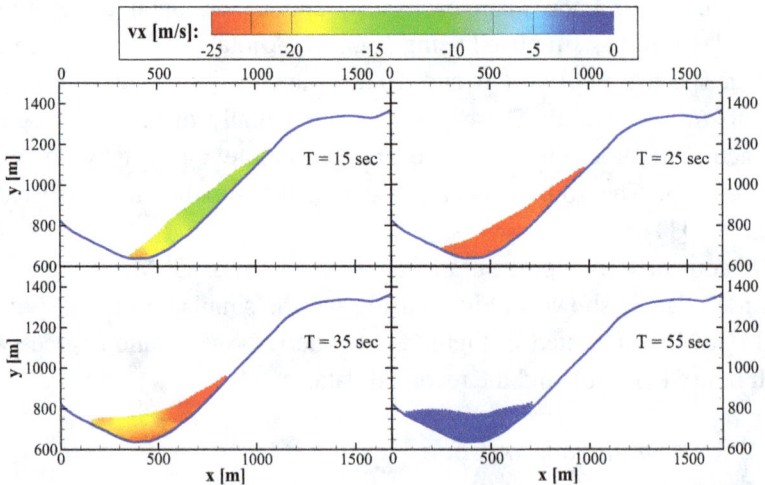

Figure 8.14. Simulated run-out process of Tangjiashan landslide at four time instances by L-GSM. v_x represents the horizontal velocity (Reprinted with permission from [6])

Table 8.5. Soil parameters used in simulating the Wangjiayan landslide [13].

Δx [m]	ρ [kg/m^3]	K [GPa]	G [GPa]	c [kPa]	ϕ [°]	ψ [°]
2	2000	1.3	0.8	30	35	0

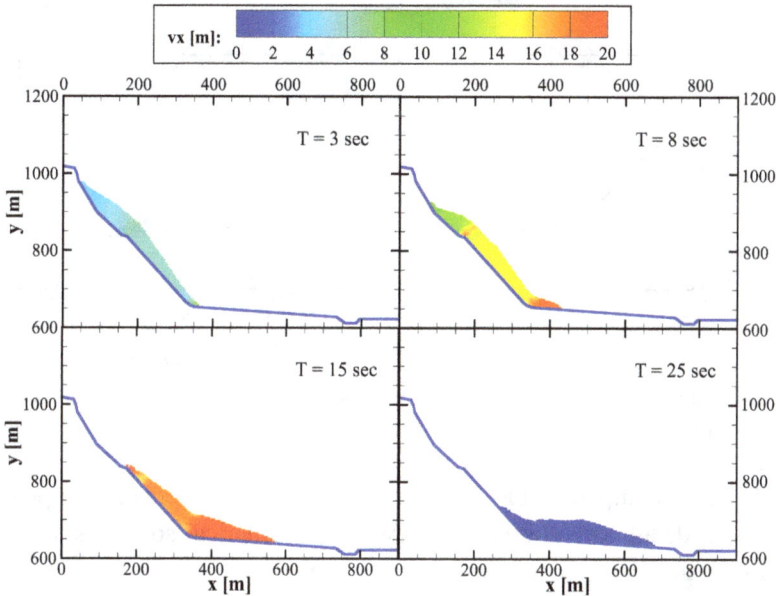

Figure 8.15. Flow process of the Wangjiayan landslide simulated by local L-GSM. v_x represents the horizontal velocity (Reprinted with permission from [6])

It had a height difference between the front and rear edge of 350 m and a sliding distance of 550 m [12] as shown in Figure 8.13. The existence of a sudden protrusion of the sliding slope resulted in obvious tensile stress on the soil materials, making this event an ideal case to investigate the numerical stability.

The particle model had 3691 real particles with a spacing of $d = 2$ m and 1605 boundary particles with a spacing of $d/3$. The soil parameters used in this model can be found in Table 8.5.

Figure 8.15 shows the flow process obtained by local L-GSM. The final distributions of the landslide obtained from global L-GSM, local L-GSM,

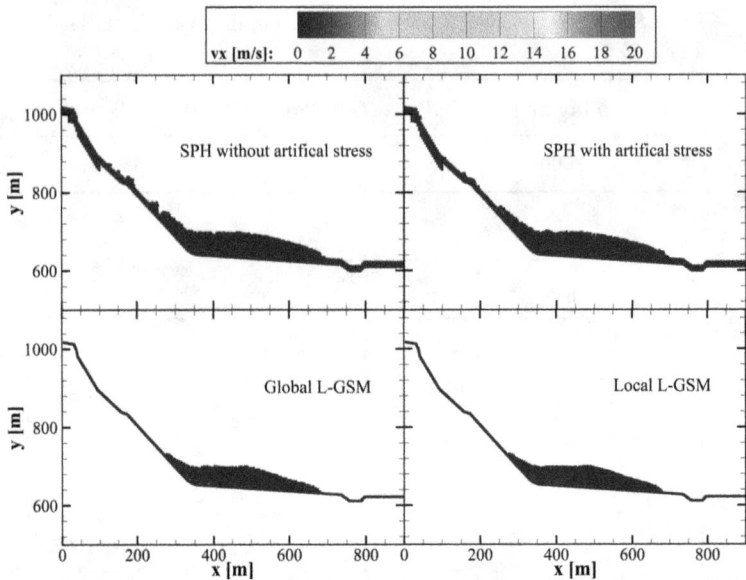

Figure 8.16. Final distribution of particles in the Wangjiayan landslide by SPH, global L-GSM, and local L-GSM, when the velocity field becomes zero (Reprinted with permission from [9]).

and SPH are compared in Figure 8.16. Figure 8.17 plots these final profiles in greater detail with zoomed-in views. The numerical solutions all agree well with the record topography.

Figure 8.18 plots zoomed-in views of the particle distributions, obtained using these three methods. It reveals that SPH particles experience obvious tensile instability exhibited as unphysical numerical cracking in areas where the soil is under tension, as shown in Figure 8.18(a). The cracking is somewhat suppressed by adopting the artificial stress remedy.

In contrast, Figure 8.18 confirms that global L-GSM and local L-GSM are both free from "tensile instability", as also demonstrated in our earlier examples.

8.3　Efficiency estimation

The computational efficiency of global L-GSM, local L-GSM, and SPH can be roughly estimated in theory based on the number of floating-point operations (FLOPs) and memory required to solve linear system equations.

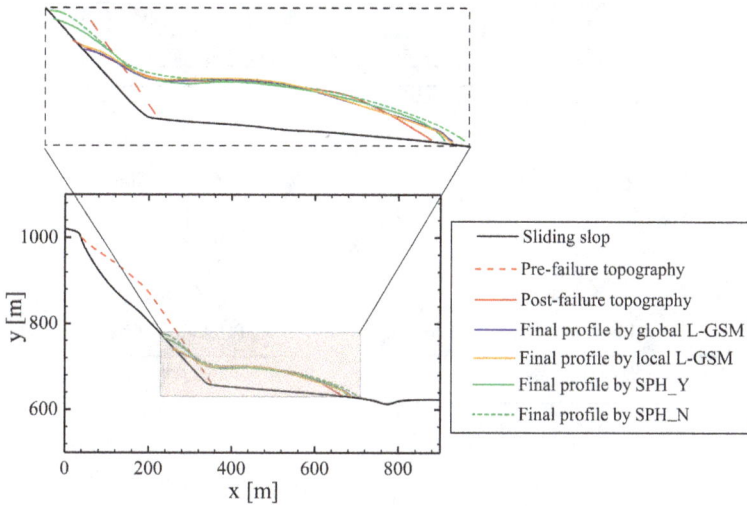

Figure 8.17. Comparison for the final profiles of the Wangjiayan landslide simulated using L-GSM and SPH, together with the onsite topography record.

Both L-GSM and SPH are explicit solvers. Thus, the discretized system equations are essentially linear at each time step, and can be expressed simply as

$$\mathbf{K}^{i+1}\mathbf{U}^i = \mathbf{U}^{i+1} \qquad (8.15)$$

where the unknown variable vector $\mathbf{U}^{i+1}$ is calculated by multiplying the coefficient matrix $\mathbf{K}^{i+1}$ obtained at the current time step with the known variable vector $\mathbf{U}^i$ obtained in the previous step. Vector $\mathbf{U}$ has a length of N_{DOF}, where DOF denotes the degree of freedom for each particle and N_{DOF} is the total number of DOFs. Depending on the type of problem, each particle can carry d field variables (including density ρ, acceleration a_x and a_y, stress tensor σ_{xx}, σ_{xy}, and σ_{yy}). When the total number of particles is N, $N_{DOF} = d \times N$. Matrix $\mathbf{K}$ has a dimension of $N_{DOF} \times N_{DOF}$, but sparse. It has a bandwidth of DOF*N_{nb}, where N_{nb} is the average number of particles supporting the gradient smoothing calculation for particles. In actual simulations, the averaged neighboring particles in SPH is 20, while in L-GSM, it is about 5.5.

Since one of the major computations is to solve the governing equation as shown in Eq. (8.12), the efficiency of L-GSM and SPH can be estimated roughly by the FLOPs needed for solving Eq. (8.12). It requires

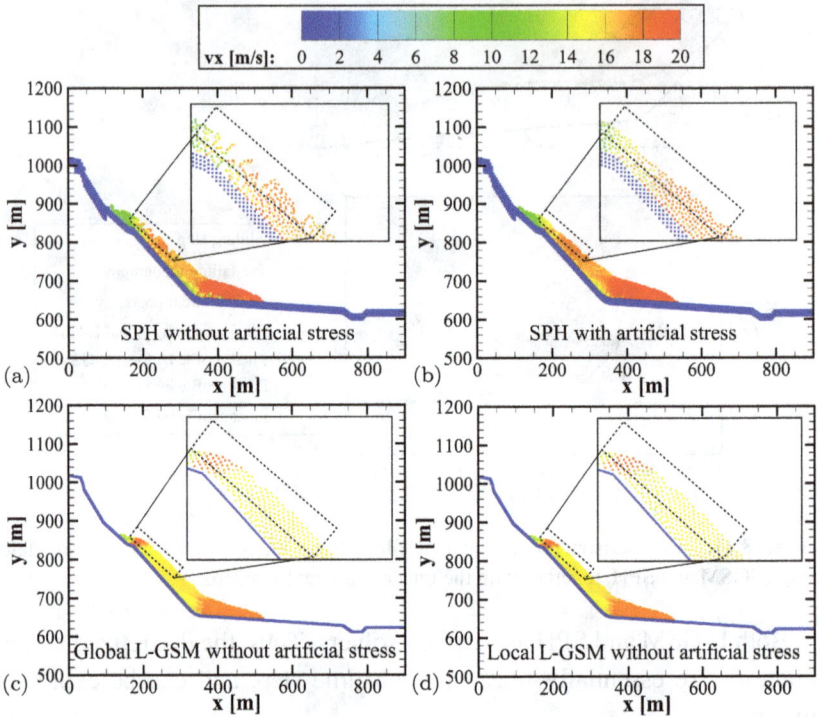

Figure 8.18. Wangjiayan landslide simulation results. Unphysical numerical cracking as the result of tensile instability of SPH without artificial stress (a), SPH with artificial stress suppressing somewhat the numerical cracking (b), global L-GSM (c) and local L-GSM (d) without any extra treatment (Reprinted with permission from [9]).

$(2\text{DOF} \times N_{\text{nb}} - 1)N_{\text{DOF}}$ FLOPs for computation, and $8 \times (\text{DOF} \times N_{\text{nb}} \times N_{\text{DOF}} + 2N_{\text{DOF}})$ bytes for memory storage. Thus, the computational complexity of solving the matrix-vector multiplication in Eq. (8.12) is of $\text{O}(N \times N_{\text{nb}})$.

Because of the difference in N_{nb} for L-GSM and SPH, the efficiency will be different. The ratio of the required number of FLOPs in L-GSM and SPH should be $\approx 0.26 (= 5.5/21)$. This is a baseline for comparison in solving the system equation at a time step. A similar ratio also applies to storage.

In actual computations, the overall calculation process involves three major parts: the gradient approximation, solving the discretized governing equations, and other time-consuming processes, as shown in Figure 8.19.

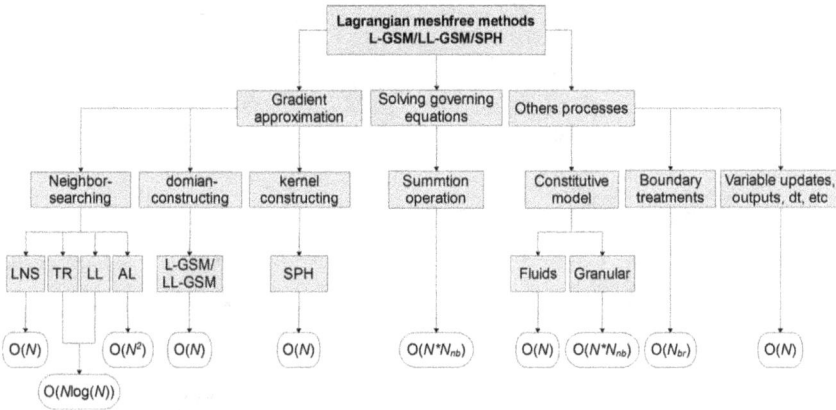

Figure 8.19. Computational complexity of various operations needed in an explicit 2D numerical model. N: total number of particles; LNS: local neighbor-searching algorithm; TR: Delaunay triangulation algorithm; LL: link-list algorithm; AL: all-list algorithm. N_{br}: number of supporting particles (Reprinted with permission from [9]).

The gradient approximating process includes the particle search, the domain construction (L-GSM) or kernel function calculation (SPH), and the summation operations for gradient calculation. This breakdown analysis of computational complexity shows that the dominant cost may not be the major one.

Numerical tests have been conducted to examine the actual computational efficiency of global L-GSM, local L-GSM, and SPH applied to hydrodynamics problems.

8.4 Efficiency via numerical tests

The efficiency of these three numerical schemes is compared for various particle spacings $\Delta x = 5.0/2.5/2.0/1.25/1.0/0.5 \times 10^{-5}$ m. Reducing particle spacing results in an increase in the number of particles N.

The numerical model used is explained in Section 7.2.1. The results are presented in Figure 8.20. The comparison shows that, with a large number of particles, local L-GSM is more efficient than global L-GSM, which in turn is more efficient than SPH. However, when the number of particles is less than 5000, SPH is more advantageous. This is because of the increasing proportion of computation time spent on the construction of the

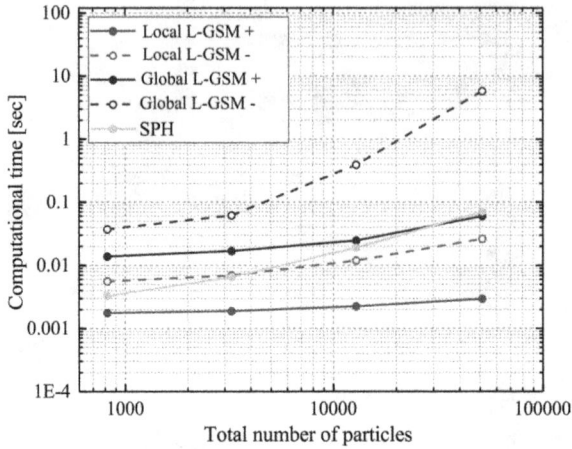

Figure 8.20. Efficiency comparison of local L-GSM, global L-GSM, and SPH schemes in hydrodynamics applications. For global L-GSM and local L-GSM, "+" means particle connection is updated by every 20 time steps, while "−" means that it is updated at every time step. When most dense particles are used, both local L-GSM and global L-GSM can be much more efficient than SPH (Reprinted with permission from [9]).

smoothing domain in global L-GSM and local L-GSM. The effect of a lower updating frequency of particle connection on the computational efficiency of L-GSM is also investigated and shown to significantly enhance their efficiency. The updating frequency used here is once every 20 steps.

8.5   Concluding remarks

The theoretical and numerical studies presented in Chapters 6–8 reveal several key findings:

- First, the global L-GSM and local L-GSM can efficiently handle large deformation problems in hydrodynamics and geotechnical problems with moving free surfaces.
- Second, in terms of accuracy, the global L-GSM and local L-GSM are at least 1st order accurate, which can be further improved through special corrections. These corrected forms of the GSM gradient operator ensure good accuracy even under extreme particle distribution conditions. In particular, L-GSM can produce error-free gradient approximation for arbitrary linear functions.

- Third, the global L-GSM and local L-GSM are proven and evidenced to be free from the tensile instability problems, which are associated with SPH.
- Fourth, the computational efficiency of the global L-GSM and local L-GSM is superior to SPH due to the use of fewer supporting particles for gradient approximation and less frequent updates of particles' connections.

Note that, the global L-GSM ensures the conservation of mass, momentum, and energy. Constructing gradient smoothing domains for the global L-GSM demands significant effort. The computational efficiency of global L-GSM can be greatly improved by developing adaptive remeshing algorithms.

The local L-GSM is relatively easy, but still more complicated than SPH. The gradient approximation in SPH does not need meshing. The local L-GSM is in general not conservative. Also, local L-GSM only approximates the conservation of momentum and energy but can produce quite accurate numerical solutions unless using a fine enough resolution of particle model and sufficiently small marching time step.

References

[1] H. H. Bui, R. Fukagawa, K. Sako, and S. Ohno, Lagrangian meshfree particles method (SPH) for large deformation and failure flows of geomaterial using elastic-plastic soil constitutive model, *International Journal for Numerical and Analytical Methods in Geomechanics*, 32(12), 1537–1570 (2008, August).

[2] W.-F. Chen and E. Mizuno, *Nonlinear Analysis in Soil Mechanics: Theory and Implementation*. Amsterdam: Elsevier, (1990).

[3] Z. Mao and G. R. Liu, A Lagrangian gradient smoothing method for solid-flow problems using simplicial mesh, *International Journal for Numerical Methods in Engineering*, 113(5), 858–890 (2018, February). doi: 10.1002/nme.5639.

[4] Z. Mao, G. R. Liu, and Y. Huang, A local Lagrangian gradient smoothing method for fluids and fluid-like solids: A novel particle-like method, *Engineering Analysis with Boundary Elements*, 107, 96–114 (2019, October). doi: 10.1016/j.enganabound.2019.07.003.

[5] Z. Mao and G. R. Liu, A 3D Lagrangian gradient smoothing method framework with an adaptable gradient smoothing domain-constructing algorithm for simulating large deformation free surface flows, *International Journal for Numerical Methods in Engineering*, 121(6), 1268–1296 (2020, March). doi: 10.1002/nme.6265.

[6] Z. Mao, G. Liu, Y. Huang, and Y. Bao, A conservative and consistent Lagrangian gradient smoothing method for earthquake-induced landslide simulation, *Engineering geology*, 260, (2019, October). doi: 10.1016/j.enggeo.2019.105226.

[7] G. R. Liu and M. B. Liu, *Smoothed Particle Hydrodynamics: A Meshfree Particle Method*. Singapore: World Scientific Publishing Co., (2003).

[8] Y. Zhang, J. Zhang, G. Chen, L. Zheng, and Y. Li, Effects of vertical seismic force on initiation of the Daguangbao landslide induced by the 2008 Wenchuan earthquake, *Soil Dynamics and Earthquake Engineering*, 73, 91–102 (2015). doi: 10.1016/j.soildyn.2014.06.036.

[9] Z. Mao, A novel Lagrangian gradient smoothing method for fluids and flowing solids, University of Cincinnati, (2019). Accessed: (2021, January 4). Available: http://rave.ohiolink.edu/etdc/view?acc_num=ucin1553252214052311.

[10] W. X. HU Xiewen, HUANG Runqiu, SHI Yubing, LU Xiaoping, ZHU Haiyong, Analysis of blocking river mechanism of Tangjiashan landslide and dam-breaking mode of its barrier dam, *Chinese Journal of Rock Mechanics and Engineering*, 28(1), 181–189 (2009). Accessed: (2018, August 10). Available: http://en.cnki.com.cn/Article_en/CJFDTOTAL-YSLX200901027.htm.

[11] G. Luo, X. Hu, C. Gu, and Y. Wang, Numerical simulations of kinetic formation mechanism of Tangjiashan landslide, *Journal of Rock Mechanics and Geotechnical Engineering*, 4(2), 149–159 (2012). doi: 10.3724/SP.J.1235.2012.00149.

[12] Y. Yin, F. Wang, and P. Sun, Landslide hazards triggered by the 2008 Wenchuan earthquake, Sichuan, China, *Landslides*, 6(2), 139–151 (2009). doi: 10.1007/s10346-009-0148-5.

[13] Y. Huang, W. Zhang, Q. Xu, P. Xie, and L. Hao, Run-out analysis of flow-like landslides triggered by the Ms 8.0 2008 Wenchuan earthquake using smoothed particle hydrodynamics, *Landslides*, 9(2), 275–283 (2012).

Chapter 9

Programming with GSMs and Source Codes

Contents

This chapter discusses major issues in programming with GSM. We provide the Matlab source code of Gradient Smoothing Methods (GSMs) for approximating the gradient and solving dynamics problems governed by PDEs. Note that the codes provided are only for the purpose of proof of principle of the GSMs, and the efficiency is not the major consideration. These codes are only tested by limited examples by the authors, and no thorough tests by a 3rd party have been conducted. Interested readers may make use of the codes free of charge, but all on his/her own risk. Also, the authors are not able to provide any service to the user.

    The examples presented in this chapter cover gradient approximation of polynomial functions in 2D and 3D, incompressible Poiseuille flow, and incompressible Couette flow under both Eulerian and Lagrangian frames. The purpose of this chapter is to elucidate the key ingredients of GSM coding to solve PDEs with several simple but representative examples while more efficient FORTRAN source codes for solving more complex engineering problems are also available on github (https://github.com/maozirui/GSMs_book).

- Section 9.1 introduces the implementation procedure of the provided Matlab source code.
- Section 9.2 applies the GSM gradient operator to approximate the gradient of polynomial functions in both 2D and 3D, demonstrating how the GSM gradient operator works.
- Section 9.3 presents GSM simulation of incompressible Poiseuille flow and provides the corresponding Matlab source code. This example highlights the key components of GSM framework under the Eulerian frame by comparing it with the finite difference scheme.
- Section 9.4 introduces the key ingredients of L-GSM simulation under the Lagrangian frame.

9.1 Overview

Solving real dynamics problems governed by a PDE or a set of PDEs via GSMs consists of the following steps: (a) connecting particles/nodes (it is called node in GSM while particle in L-GSM), (b) ordering supporting nodes/particles in sequence, (c) constructing gradient smoothing domain (GSD), (d) calculating the area of GSD, (e) approximating the gradient with the GSM gradient operator, and (f) inserting the smoothed gradient into PDEs.

(A) *Particles/nodes connection*: Under an Eulerian frame, the nodes are fixed spatially during the whole dynamic process and connected using a global mesh over the entire domain, which does not need to be updated over time. In contrast, under a Lagrangian frame, the particles will move along with the deforming and moving material/media, and require the connection to be updated properly. The connection of particles in L-GSM can be attained either by constructing a global mesh as done under the Eulerian frame or by employing a local grid. The attached code employs a local grid.

(B) *Ordering supporting particles*: The construction of GSD relies on properly ordered supporting nodes/particles. Under an Eulerian frame, the sequence of connecting nodes has been arranged automatically

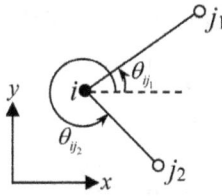

Figure 9.1. Relative direction of supporting particle from target particle.

when using the built-in "Triangulation" function in Matlab. Under a Lagrangian frame, as a local grid is employed in L-GSM, the ordering must be done manually based on the direction of each supporting node/particle relative to the target node/particle (see Figure 9.1):

$$\theta_{ij_k} = \begin{cases} \arccos\left(\frac{x_{j_k}-x_i}{r_{ij}}\right), y_{j_k} \geq y_i \\ 2\pi - \arccos\left(\frac{x_{j_k}-x_i}{r_{ij_k}}\right), y_{j_k} < y_i \end{cases} \tag{9.1}$$

(C) *Construction of GSD*: In the provided code, the node-based GSD or n-GSD is employed. The n-GSD for a target node/particle is constructed by sequentially connecting the midpoints of edges and the centroids of triangles that connect the supporting particles with the target particle.

(D) *Area calculation of n-GSD*: The area of the n-GSD is calculated as 1/3 of the polygon formed by the supporting particle. The polygon's area can be calculated by summing up the areas of each triangle composed by each pair of adjacent supporting node/particle and the target node/particle. The area of each triangle is computed using the following formula:

$$Area_{\Delta ijk} = \frac{(x_j - x_i)(y_k - y_i) - (x_k - x_i)(y_j - y_i)}{2} \tag{9.2}$$

(E) *Gradient approximation*: Based on the gradient smoothing theory introduced in Chapter 6, the gradient of a field variable U has the simplified form of

$$\begin{cases} \frac{\partial U_i}{\partial x} = \frac{1}{6V_i}\sum_{k=1}^{n_i}(U_i+U_{j_k})(y_{j_{k+1}}-y_{j_{k-1}}), \\ \frac{\partial U_i}{\partial y} = \frac{1}{6V_i}\sum_{k=1}^{n_i}(U_i+U_{j_k})(x_{j_{k+1}}-x_{j_{k-1}}). \end{cases} \tag{9.3}$$

(F) After obtaining the gradients of all necessary field variables, the smoothed gradients are inserted into the governing PDEs as done in

FDM and SPH, and the increment of field variables within a properly small time interval can be computed from the governing equations for time integration.

9.2 GSM gradient approximation

9.2.1 *2D cases*

In this subsection, we demonstrate the usage of the provided Matlab source code for the 2D GSM gradient operator by approximating the gradient of a simple polynomial function, $F(x, y) = x^3$. The source code allows users to test the accuracy of the GSM gradient operator under various distribution conditions of supporting nodes/particles, including uniform distribution, slightly irregular distribution, severely irregular distribution, and boundary nodes/particles, as shown in Figure 9.2. The normalized GSM gradient operator presented in Section 6.3.2 is used to address the particle deficiency issue of boundary nodes/particles.

Use of source code: The attached source code enables users to perform the following tests: (a) testing the accuracy of the GSM gradient operator under

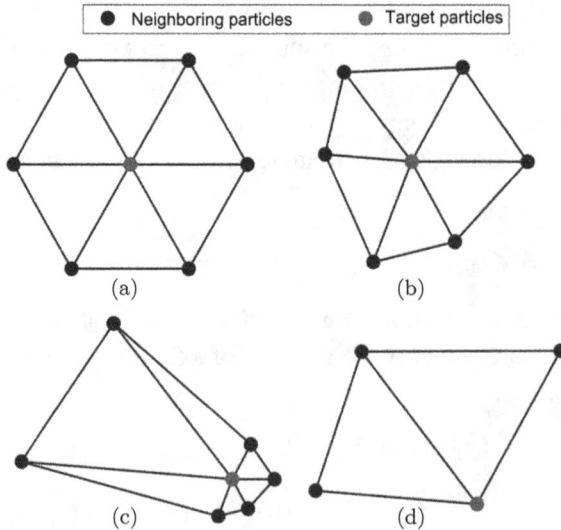

Figure 9.2. Four distribution cases of supporting particles considered in the source code.

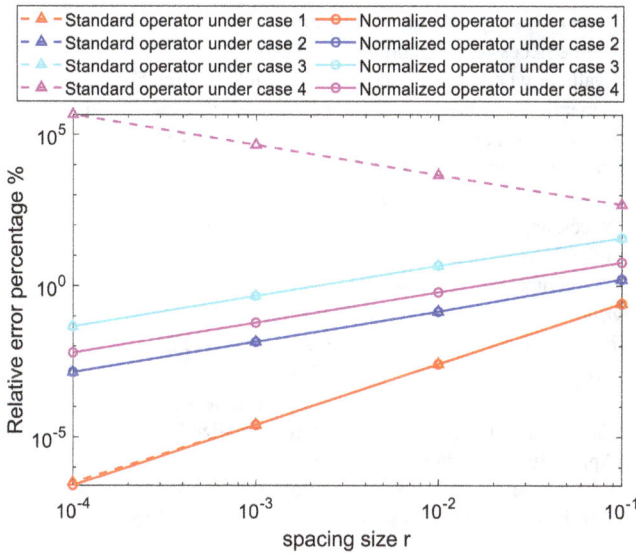

Figure 9.3. Accuracy order of standard GSM operator and normalized operator under the 4 given conditions of particles distributions.

different distribution conditions of supporting nodes/particles by switching the type_support in line 7 of the main code named as "GSM_2D.m"; (b) comparing the performance of the standard GSM gradient operator and the normalized one in accuracy of gradient approximation near the boundary (by setting "type_support = 4"); (c) testing the accuracy condition of the GSM gradient operator under different particles spacings by changing the values of "r" in line 12 of the Main code "GSM_2D".

The testing results of the source code are presented in Figure 9.3. Users are free to test other types of functions by changing the "f_ref" in line 13 and "f" in line 40 (if the testing function is changed, remember to change the exact solution in line 14 used for error calculation).

Structure of source code

Outline of main program 'GSM_2D.m'

Necessary comments and initialization % Line 1–4
Define necessary controlling parameters and functions % Line 5–14
Generate supporting particles % Line 15–41
Plot out the supporting particles by calling 'plot particles' % Line 42

The source code including the main program "GSM_2D.m" and the subfunctions of "plot_particles.m", "GSM_standard.m", and "GSM_normalized.m" are presented as follows.

Main program: GSM_2D.m

This program approximates the gradient of a polynomial function $F = x^3$ with standard and normalized GSM gradient operators under a 2D case. The distribution of supporting particles is plotted out, and the numerical error of GSM solutions is computed and printed out.

  The program makes calls on the following functions:

- **plot_particles.m**
- **GSM_standard.m**
- **GSM_normalized.m**

Listing:

```
1   % This code is written to approximate gradient of scalar
    function F
2   % using GSM gradient operator. For detailed information,
3   % please refer to "Zirui Mao, G.R. Liu, Int J Numer Methods Eng.
    2018;113:858—C890.".
4   clc; clear all; clf;
5   %%%%%%% Controlling parameters %%%%%%%%%
6   o=[1.0 1.0]; % 2D coordinate of target node/particle
7   type_support=2; % type switch of supporting particles
```

```
8                        % = 1 uniformly distributed supporting particles
9                        % = 2 irregular distribution with small distance
        variation
10                       % = 3 irregular distribution with large distance
        variation
11                       % = 4 boundary particle with particle deficiency
12   r=0.001; % define the spacing size
13   f_ref=o(1)^3; % define the testing scalar function F
14   dfx_ref=3*o(1)^2; % give the exact solution of F's gradient
15   %%%%%%% create types of supporting particles %%%%%%%%%%%%%%%
16   if type_support==1 % uniformly distributed supporting particles
17       N=6; % amount of supporting particles
18       factor_length=[1 1 1 1 1 1]; % normalized length of edges
19       factor_angle= [1 1 1 1 1 1]; % normzlied angles between each
     pair of adjacent supporting particles
20   elseif type_support==2 % irregular distribution with small
     distance variation
21       N=6;
22       factor_length=[1 1.05 0.83 1.0 0.85 1.1];
23       factor_angle=[1 1.1 0.78 1.2 0.83 1];
24   elseif type_support==3 %irregular distribution with large distance
        variation
25       N=6;
26       factor_length=[1 5 5.5 1.0 0.85 1.1];
27       factor_angle=[1 1.1 0.78 1.2 0.83 1];
28   elseif type_support==4 % boundary particle with particle
     deficiency
29       N=3;
30       factor_length=[1 1.1 0.95 1.0 0.85 1.1];
31       factor_angle=[1 1.1 0.78 1.2 0.83 1];
32   end
33   x=zeros(N,1); y=zeros(N,1); Length=r*factor_length; angle=zeros
     (N,1);
34   f=zeros(N,1); % scalar function F
35   %%%%%%% generate coordinates of supporting particles
     %%%%%%%%%%%%%%
36   for i=1:N
37       angle(i)=360/sum(factor_angle)*(sum(factor_angle(1:i)));
38       x(i)=o(1)+Length(i)*cosd(angle(i)); % specify the x
     coordinates of supporting particles
39       y(i)=o(2)+Length(i)*sind(angle(i)); % specify the y
     coordinates of supporting particles
40       f(i)=x(i).^3; % define the function values of supporting
     particles
41   end
```

```
42  plot_particles( N,x,y,o,angle);%%%%%%%%%% plot the particles
    configuration
43  %%%%%%%% Gradient approximation with GSM gradient operator
    %%%%%%%%%%%%%%%%
44  [dfx_1] = GSM_standard( N,x,y,o,angle,f,f_ref); % calculate
    gradient with standard GSM operator
45  [dfx_2] = GSM_normalized( N,x,y,o,angle,f,f_ref); % calculate
    gradient with normalized GSM operator
46  error1=abs(dfx_1-dfx_ref)/dfx_ref*100; error2=abs(dfx_2-dfx_ref)/
    dfx_ref*100; % calculate relative error percentage
47  %%%%%%%% print out the solutions and compare to the exact
    solution %%%%%%
48  fprintf('Standard GSM operator = %0.2f | Exact solution = %0.2f |
    Error percentage of Standard GSM operator = %0.2e \n', dfx_1,
    dfx_ref,error1);
49  fprintf('Normalized GSM operator = %0.2f | Exact solution = %0.2f
    | Error percentage of Normalized GSM operator = %0.2e \n', dfx_2,
    dfx_ref, error2);
```

plot_particles.m

This function plots the locations of supporting particles relative to the target particle and the connection of particles.

Listing

```
1   function [] = plot_particles(N,x,y,o,angle)
2   X=x; Y=y;
3   XX=X;YY=Y;
4   k=0;
5   %%%%%%%% evaluate whether boundary particle %%%%%%%%%%%
6   for j=1:N
7       jp1=j+1; % index of next particle in the anti-clock direction
8       if j==N
9           jp1=1;
10      end
11      angle_up=angle(jp1)-angle(j); % included angle of two adjacent
    supporting particles
12      if angle_up<0
13          angle_up=angle_up+360;
14      end
15      if angle_up>180 % boundary particle
16          k=j;
17      end
18  end
19  %%%% create the particles connecting lines %%%%%%%%
```

```
20   if k==0
21       X=[x; x(1)]; Y=[y; y(1)];
22   else
23       if k~=N % boundary particle
24           j=k;
25           for i=1:(N-j)
26               X(i)=XX(j+i); Y(i)=YY(j+i);
27           end
28           for i=1:j
29               X(N-j+i)=XX(i); Y(N-j+i)=YY(i);
30           end
31       end
32   end
33   Ox=[o(1); x(1)]; Oy=[o(2); y(1)];
34   for i=2:N
35       Ox=[Ox;o(1);x(i)]; Oy=[Oy;o(2);y(i)];
36   end
37   %%%%%%%%% plot the particles and their connection %%%%%%%%%%
38   H=zeros(1,4);
39   set(0,'defaultfigurecolor','w')    % set the background as white.
40   H(1:2)=plot(X,Y,'b-',Ox,Oy,'b-','linewidth',1.5);
41   hold on;
42   H(3)=plot(x,y,'bo','MarkerFaceColor','b','markersize',10);
43   hold on;
44   H(4)=plot(o(1),o(2),'ro','MarkerFaceColor','r','markersize',10);
45   legend(H([3 4]),'Neighboring particles','Target particles');
46   axis equal
47   xlabel('x'); ylabel('y');
48   set(gca,'FontName','TimesRoman','FontSize',14)
```

GSM_standard.m

This function approximates the gradient of function f with the standard GSM gradient operator.

Listing:

```
1    function [dfx] = GSM_standard( N,x,y,o,angle,f,f_ref)
2    % This function is written to approximate gradient of scalar
     function
3    % f with the standard GSM gradient operator based on n-GSD.
4    dfx=0;
5    %%%%%%%%%%[start] area calculation [start] %%%%%%%%%%%%%%
6    area=(x(N).*y(1)-x(1).*y(N))/6;
7    for j=1:N-1
8        area=area+(x(j).*y(j+1)-x(j+1).*y(j))/6; % calculate the area
     V of n-GSD
```

```
9   end
10  k=0;
11  %%% evaluate the target particle is on boundary or not
12  for j=1:N
13      jp1=j+1;
14      if j==N
15          jp1=1;
16      end
17      angle_up=angle(jp1)-angle(j);
18      if angle_up<0
19          angle_up=angle_up+360;
20      end
21      if angle_up>180
22          k=j;
23      end
24  end
25  if k~=0 % meaning the target particle locates on boundary
26      X=x; Y=y;
27      if k~=N
28          j=k;
29          for i=1:(N-j)
30              X(i)=x(j+i); Y(i)=y(j+i);
31          end
32          for i=1:j
33              X(N-j+i)=x(i); Y(N-j+i)=y(i);
34          end
35      end
36      X=[X; o(1)]; Y=[Y; o(2)];
37      area=(X(N+1).*Y(1)-X(1).*Y(N+1))/6; % recalculate the area of
    n-GSD for boundary particle
38      for j=1:N
39          area=area+(X(j).*Y(j+1)-X(j+1).*Y(j))/6;
40      end
41  end
42  %%%%%%%%%%%%%% gradient approximation %%%%%%%%%%%%%%%%
43  for j=1:N
44      jm1=j-1;jp1=j+1;
45      if j==1
46          jm1=N;
47      elseif j==N
48          jp1=1;
49      end
50      angle_up=angle(jp1)-angle(j); % Left part
51      if angle_up<0
52          angle_up=angle_up+360;
```

```
53      end
54      angle_low=angle(j)—angle(jm1); % right part
55      if angle_low<0
56          angle_low=angle_low+360;
57      end
58      if angle_up<180
59          x_up=(x(jp1)+x(j)+o(1))/3; y_up=(y(jp1)+y(j)+o(2))/3;
60          dSx=y_up—(o(2)+y(j))/2;
61          dfx=dfx+(f(j)+f_ref)/2.*dSx/area;
62      end
63      if angle_low<180
64          x_low=(x(jm1)+x(j)+o(1))/3; y_low=(y(jm1)+y(j)+o(2))/3;
65          dSx=—y_low+(o(2)+y(j))/2;
66          dfx=dfx+(f(j)+f_ref)/2.*dSx/area;
67      end
68   end
69   end
```

GSM_normalized.m

This function approximates the gradient of function f with the normalized GSM gradient operator.

```
1    function [dfx] = GSM_normalized( N,x,y,o,angle,f,f_ref)
2    % This function is written to approximate gradient of scalar
     function f
3    % with the normalized GSM gradient operator. For detailed
     information,
4    % please refer to "Zirui Mao., et al.,Int J Numer Methods Eng.
     2018;113:858—C890."
5    sigma1=0; sigma2=0; sigma3=0; sigma4=0; A=0; B=0;  % necessary
     parameters
6    for j=1:N
7        jm1=j—1;jp1=j+1; % jm1 = last supporting particle; jp1 = next
     particle
8        if j==1
9            jm1=N;
10       elseif j==N
11           jp1=1;
12       end
13       %%%%% calculate n*dS in the GSM gradient operator %%%%%%%%%
14       angle_up=angle(jp1)—angle(j); % angle from next supporting
     particle
15       if angle_up<0
16           angle_up=angle_up+360;
17       end
```

```
18      angle_low=angle(j)—angle(jm1); % angle from last supporting
    particle
19      if angle_low<0
20          angle_low=angle_low+360;
21      end
22      if angle_up>180 % boundary particle
23          x_up=(o(1)+x(j))/2; y_up=(o(2)+y(j))/2;
24      else
25          x_up=(x(jp1)+x(j)+o(1))/3;   y_up=(y(jp1)+y(j)+o(2))/3;
26      end
27      if angle_low>180 % boundary particle
28          x_low=(o(1)+x(j))/2 ;  y_low=(o(2)+y(j))/2;
29      else
30          x_low=(x(jm1)+x(j)+o(1))/3;   y_low=(y(jm1)+y(j)+o(2))/3;
31      end
32      dSx=y_up—y_low; dSy=x_up—x_low;
33      %%%%%% calculate sigmas, A and B in the normalized GSM
    operator
34      sigma1=sigma1+dSx.*(x(j)—o(1));      sigma2=sigma2+dSx.*
        (y(j)—o(2));
35      sigma3=sigma3+dSy.*(x(j)—o(1));      sigma4=sigma4+dSy.*
        (y(j)—o(2));
36      A=A+dSx.*(f(j)—f_ref);       B=B+dSy.*(f(j)—f_ref);
37  end
38  dfx=(sigma4.*A—sigma2.*B)./(sigma1.*sigma4—sigma2.*sigma3); %
    normalized GSM operator
39  end
```

9.2.2 *3D cases*

In this section, we showcase how the 3D GSM operator approximates the gradient of a simple polynomial function, $F(x, y, z) = x^2/2 + y^3/3 + z^4/4 + 10(x + y + z)$. The main challenge in this process lies in constructing the 3D n-GSD, as discussed in Section 6.5.2. The provided source code covers three types of GSD: regular box, irregular convex–concave geometry, and open box with boundary particles experiencing particle deficiency beyond the boundary (see Figure 9.4). The gradient is approximated with the 3D normalized GSM gradient operator, as described in Section 6.3.2.

Use of source code: The source code enables users to conduct the following explorations: (a) verify the accuracy of the 3D GSM gradient

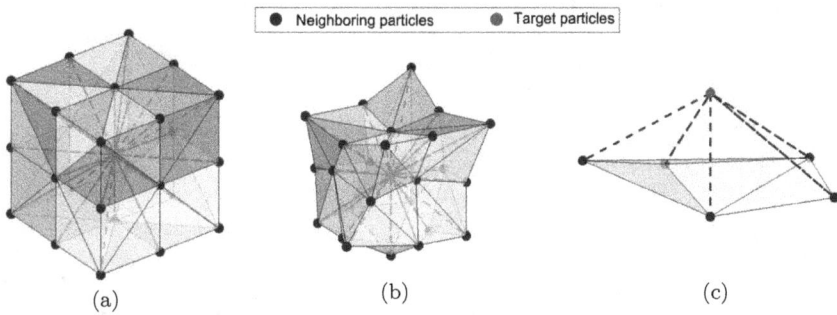

Figure 9.4. Three distribution conditions of supporting particles considered in the source code.

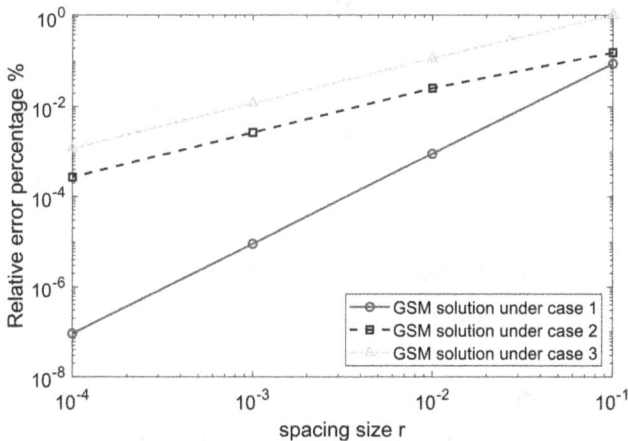

Figure 9.5. Accuracy condition of 3D normalized GSM operator under the 3 given distribution conditions of supporting particles.

approximation; (b) examine the effectiveness of the 3D normalized operator under different distribution conditions of supporting nodes/particles by switching the "type_support" in line 7 of the main code "GSM_3D.m"; (c) evaluate the accuracy condition under different spacing sizes by changing the factor "r" in line 11.

Some testing results have been plotted in Figure 9.5. Similarly, users are also free to test other functions by changing the "f" in line 12 and "f_neighbor" in line 34 of the main code "GSM_3D.m". Remember to change the related exact solution for error calculation in line 13 as well.

Structure of source code

Outline of main program 'GSM_3D.m'

> Necessary comments and initialization % Line 1–4
> Define necessary controlling parameters and functions % Line 5–13
> Generate supporting particles % 14–28
> Construct 3D n-GSM for GSM gradient operator % Line 29–30
>> *Outline of 'GSD_construction.m'* at Line 30 (see Fig. 4.10 for details)
>>> Necessary comments and variable definitions % Line 2–7
>>> Step 1 in Fig. 4.10: normalize relative distance % L 8–14
>>> Step 2 in Fig. 4.10 : find the surface triangles % Line 15–269
>>>> Step 2.1: find the closest two supporting particle i1, i2 % Line 16–24
>>>> Step 2.2: find the first surface triangle % Line 25–108
>>>> Step 2.3: find the subsequent surface triangles % Line 109–269
>>> Calculate n*dS in the GSM operator % Line 270–281
>>> Plot supporting particles in 3D % 282–290
> Approximate gradient with 3D normalized GSM operator % Line 31–38
> Calculate numerical error of GSM solution % Line 39
> Print out GSM solutions and numerical errors % Line 40–42

The source code including the main program "GSM_3D.m" and the function of "GSD_construction.m" is presented as follows.

Main program: GSM_3D.m

This program approximates the gradient of a polynomial function $F = x^2/2 + y^3/3 + z^4/4 + 10(x + y + z)$ with a 3D normalized GSM gradient operator. The distribution of the supporting particles is plotted out, and the numerical error of GSM solutions is computed and printed out.

The program makes calls to the following functions:

- **GSD_construction.m**

Listing

```
1   % This code is written to approximate gradient of scalar F in 3D
    case
2   % using 3D normalized GSM gradient operator. For detailed
    information, please refer to
3   % "Zirui Mao, et al., Int J Numer Methods Eng. 2020. vol. 121,
        no. 6, pp. 1268–C1296."
4   clc; clear all; clf;
5   %%%%%%% controlling parameters %%%%%%%%
6   o1=1; o2=1; o3=1; % define 3D coordinates of target particle
```

```
7   type_support=1; % type switch of supporting particles
8                     % = 1 regular box
9                     % = 2 mixed convex—concave domain
10                    % = 3 corner boundary particle
11  r=0.01; % define the spacing size
12  f=o1^2/2+o2^3/3+o3^4/4+10*(o1+o2+o3); % testing scalar function F
13  dfx_theory=o1+10; dfy_theory=o2^2+10; dfz_theory=o3^3+10; %
    gradient of F in theory
14  %%%%%% generate positions of neighbors relative to  target
    particle %%%%%%%
15  if type_support == 1 % regular box
16      x=[0; 1; 1.; 0; −1; −1.; −1; 0; 1; 1; 1; 0; −1.; −1; −1; 0; 1;
    0; 1; 1; 0; −1; −1.; −1; 0; 1];
17      y=[0; 0; 1.; 1; 1; 0; −1; −1; −1; 0; 1; 1.; 1; 0; −1; −1; −1;
    0; 0; 1; 1; 1.; 0; −1; −1.; −1];
18      z=[−1; −1.;  −1;  −1; −1; −1; −1.; −1; −1; 0; 0; 0; 0; 0; 0;
    0; 0; 1.; 1; 1; 1.; 1; 1; 1; 1; 1];
19  elseif type_support == 2 % mixed convex—concave domain
20      x=[0; 1.5; 0.9; 0; −1; −1.; −1; 0; 1; 0.8; 1; 0; −1.; −0.9;
    −1; −0.2; 1; 0; 1; 1; 0; −1; −1.5; −1; 0; 1];
21      y=[0; 0; 1.; 1; 1.5; 0; −1; −0.8; −1; 0; 1; 1.2; 1; 0; −1; −1;
    −1; 0; 0; 1.2; 1; 0.7; 0; −1.5; −1.; −1];
22      z=[−1.5; −1.1; −1; −1; −0.75; −1; −1.; −1.2; −1; 0; 0; 0.3; 0;
    0.2; 0; 0; −0.3; 0.8; 1; 1; 0.75; 1; 1.2; 1; 1; 1.5];
23  elseif type_support == 3 % corner
24      x=[0;    1;   1.25;   0;   −0.75];
25      y=[0;    0;   1.25;   1;   0.75];
26      z=[−1; −1.;  −1.25;  −1; −0.75];
27  end
28  x=x*r; y=y*r; z=z*r; % generate the relative position of
    supporting particles
29  %%%%%%%%%%%%%% 3D n—GSD construction %%%%%%%%%%%%%%
30  [dsx,dsy,dsz] = GSD_construction (x, y, z); % output dS_x, dS_y,
    and dS_z in GSM operator
31  % %%%% Approximation of gradient with the 3D normalized GSM
    operation %%%%%%%%%%%%
32  dfx=0; dfy=0; dfz=0; % gradient of f in x, y, and z directions,
    respectively
33  for k=1:length(x)
34  f_neighbor=(o1+x(k))^2/2+(o2+y(k))^3/3+(o3+z(k))^4/4+10*(o1+x(k)+
    o2+y(k)+o3+z(k));
35      dfx=dfx+(f_neighbor—f)*dsx(k); % x componenet of gradient of F
36      dfy=dfy+(f_neighbor—f)*dsy(k); % y componenet of gradient of F
37      dfz=dfz+(f_neighbor—f)*dsz(k); % z componenet of gradient of F
38  end
```

```
39   error=abs(dfz–dfz_theory)/dfz_theory*100; % calculate relative
     error percentage
40   fprintf('dF_x: GSM solution = %0.2f | Exact solution = %0.2f \n',
     dfx,dfx_theory);
41   fprintf('dF_y: GSM solution = %0.2f | Exact solution = %0.2f \n',
     dfy,dfy_theory);
42   fprintf('dF_z: GSM solution = %0.2f | Exact solution = %0.2f |
     Error percentage of GSM solution = %0.2e \n', dfz, dfz_theory,
     error);
```

GSD_construction.m

This function constructs the necessary 3D n-GSD for the GSM gradient operator, calculates the corresponding term n*dS in the GSM operator, and plots the relative locations of supporting particles from the target particle.

```
1    function [dsx,dsy,dsz] = GSD_construction (xx, yy, zz)
2    % This function is written to construct 3D n–GSD structure for 3D
     L–GSM.
3    % For more detail, please refer to ""Zirui Mao, et al., Int J
     Numer Methods Eng.
4    % 2020. vol. 121, no. 6, pp. 1268–C1296.""
5    X=xx; Y=yy; Z=zz;
6    N=length(X); % total number of neighbors
7    x0=X; y0=Y; z0=Z;
8    %%%%%%%%%%%%%% step 1: normalized the distances %%%%%%%%%%%%
     %%%%%
9    for i=1:N
10       l(i)=sqrt(X(i)^2+Y(i)^2+Z(i)^2);
11       x(i)=X(i)/l(i); y(i)=Y(i)/l(i);z(i)=Z(i)/l(i);
12   end
13   xt=0; yt=0; zt=0; % coordinate of target particle
14   NS=0; NE=0;
15   %%%%%%% step 2:  find the first surface triangle
     %%%%%%%%%%%%%%%%%%%
16   i1=1; min_dis=1000;
17   for i2=2:N
18       distance=sqrt((x(i1)–x(i2))^2+(y(i1)–y(i2))^2+(z(i1)–z(i2))
     ^2);
19       if distance<min_dis
20           min_dis=distance;
21           ii=i2;
22       end
23   end
24   i2=ii;
25   V=0; surf_area=zeros(N,3);
26   if N>3
```

```
27        ind=0;
28        for i=2:N—1
29            if i>=ind
30                sig=1;
31                A0=[0 0 0 1; x(i1) y(i1) z(i1) 1; x(i2) y(i2) z(i2) 1;
    x(i) y(i) z(i) 1];
32                V0=det(A0)/6;
33                for j=i+1:N
34                    A1=[x(j) y(j) z(j) 1; x(i1) y(i1) z(i1) 1; x(i2)
    y(i2) z(i2) 1; x(i) y(i) z(i) 1];
35                    V1=det(A1)/6;
36                    if V0*V1<0
37                        sig=—1; ind=j;
38                        break
39                    end
40                end
41                if sig>0&&abs(V0)>0
42                    break;
43                end
44            end
45            if i==N—1
46                i=N;
47            end
48        end
49    else
50        i1=1;i2=2;i=3;
51        A0=[0 0 0 1; x(i1) y(i1) z(i1) 1; x(i2) y(i2) z(i2) 1; x(i)
    y(i) z(i) 1];
52        V0=det(A0)/6;
53    end
54    if abs(V0)>0 % is triangle
55        NS=1; % number of surface
56        surface(NS, 1:3)=[i1 i2 i];
57        NE=3;
58        A=[0 0 0 1; x0(i1) y0(i1) z0(i1) 1; x0(i2) y0(i2) z0(i2) 1;
    x0(i) y0(i) z0(i) 1];
59        V=V+abs(det(A))/24;
60        x11=x0(i1)/2; y11=y0(i1)/2; z11=z0(i1)/2;
61        x12=(x0(i1)+x0(i2))/3; y12=(y0(i1)+y0(i2))/3; z12=(z0(i1)+
    z0(i2))/3;
62        x13=(x0(i1)+x0(i)+x0(i2))/4; y13=(y0(i1)+y0(i)+y0(i2))/4; z13
    =(z0(i1)+z0(i)+z0(i2))/4;
63        x21=x0(i2)/2; y21=y0(i2)/2; z21=z0(i2)/2;
64        x22=(x0(i2)+x0(i))/3; y22=(y0(i2)+y0(i))/3; z22=(z0(i2)+
    z0(i))/3;
```

```
65      x23=(x0(i2)+x0(i1)+x0(i))/4; y23=(y0(i2)+y0(i1)+y0(i))/4; z23
    =(z0(i2)+z0(i1)+z0(i))/4;
66      x31=x0(i)/2; y31=y0(i)/2; z31=z0(i)/2;
67      x32=(x0(i)+x0(i1))/3; y32=(y0(i)+y0(i1))/3; z32=(z0(i)+
    z0(i1))/3;
68      x33=(x0(i)+x0(i2)+x0(i1))/4; y33=(y0(i)+y0(i2)+y0(i1))/4;
    z33=(z0(i)+z0(i2)+z0(i1))/4;
69      a13=sqrt((x11-x12)^2+(y11-y12)^2+(z11-z12)^2);
70      a11=sqrt((x12-x13)^2+(y13-y12)^2+(z13-z12)^2);
71      a12=sqrt((x11-x13)^2+(y11-y13)^2+(z11-z13)^2);
72      a23=sqrt((x21-x22)^2+(y21-y22)^2+(z21-z22)^2);
73      a21=sqrt((x22-x23)^2+(y23-y22)^2+(z23-z22)^2);
74      a22=sqrt((x21-x23)^2+(y21-y23)^2+(z21-z23)^2);
75      a33=sqrt((x31-x32)^2+(y31-y32)^2+(z31-z32)^2);
76      a31=sqrt((x32-x33)^2+(y33-y32)^2+(z33-z32)^2);
77      a32=sqrt((x31-x33)^2+(y31-y33)^2+(z31-z33)^2);
78      p3=a31+a32+a33; p3=p3/2;  p1=(a11+a12+a13)/2;  p2=(a21+a22+
    a23)/2;
79      area1=sqrt(p1*(p1-a11)*(p1-a12)*(p1-a13))*2;
80      area2=sqrt(p2*(p2-a21)*(p2-a22)*(p2-a23))*2;
81      area3=sqrt(p3*(p3-a31)*(p3-a32)*(p3-a33))*2;
82      l1(1)=(y12-y11)*(z13-z11)-(y13-y11)*(z12-z11);
83      l1(2)=(z12-z11)*(x13-x11)-(z13-z11)*(x12-x11);
84      l1(3)=(x12-x11)*(y13-y11)-(x13-x11)*(y12-y11);
85      l2(1)=(y22-y21)*(z23-z21)-(y23-y21)*(z22-z21);
86      l2(2)=(z22-z21)*(x23-x21)-(z23-z21)*(x22-x21);
87      l2(3)=(x22-x21)*(y23-y21)-(x23-x21)*(y22-y21);
88      l3(1)=(y32-y31)*(z33-z31)-(y33-y31)*(z32-z31);
89      l3(2)=(z32-z31)*(x33-x31)-(z33-z31)*(x32-x31);
90      l3(3)=(x32-x31)*(y33-y31)-(x33-x31)*(y32-y31);
91      ll1=sqrt(l1(1)^2+l1(2)^2+l1(3)^2);
92      ll2=sqrt(l2(1)^2+l2(2)^2+l2(3)^2);
93      ll3=sqrt(l3(1)^2+l3(2)^2+l3(3)^2);
94      l1=l1/ll1; l2=l2/ll2; l3=l3/ll3;
95      if V0<0
96          edge(1,1:2)=[i1 i2]; edge(2,1:2)=[i2 i]; edge(3,1:2)=
    [i i1];
97          surf_area(i1,1:3)=surf_area(i1,1:3)+area1*l1;
98          surf_area(i2,1:3)=surf_area(i2,1:3)+area2*l2;
99          surf_area(i,1:3)=surf_area(i,1:3)+area3*l3;
100     else
101         edge(1,1:2)=[i2 i1]; edge(2,1:2)=[i i2]; edge(3,1:2)=
    [i1 i];
102         surf_area(i1,1:3)=surf_area(i1,1:3)-area1*l1;
103         surf_area(i2,1:3)=surf_area(i2,1:3)-area2*l2;
```

```
104              surf_area(i,1:3)=surf_area(i,1:3)—area3*l3;
105      end
106      edge_particle(1,1)=i; edge_particle(2,1)=i1; edge_particle
     (3,1)=i2;
107  end
108  Edge=edge; edge_used=0;
109  %%%%%%%%%%%%%% step 3: find the subsequent surface triangles
     %%%%%%%%%%%%%%%%%%
110  while ~isempty(edge)
111      i1=edge(1,1); i2=edge(1,2);
112      min_i=0;
113      for i=1:N
114          if i~=i1&&i~=i2&&i~=min_i&&i~=edge_particle(1)
115              belong_used=0;
116              for j=1:length(edge_used(:,1))
117                  if i==edge_used(j,1)&&i1==edge_used(j,2)
118                      belong_used=1;
119                      break
120                  elseif i1==edge_used(j,1)&&i==edge_used(j,2)
121                      belong_used=1;
122                      break
123                  elseif i==edge_used(j,1)&&i2==edge_used(j,2)
124                      belong_used=1;
125                      break
126                  elseif i2==edge_used(j,1)&&i==edge_used(j,2)
127                      belong_used=1;
128                      break
129                  end
130              end
131              if belong_used==0
132                  if min_i==0
133                      min_i=i;
134                  else
135                      A0=[x(i) y(i) z(i) 1; x(i1) y(i1) z(i1) 1;
     x(i2) y(i2) z(i2) 1; x(min_i) y(min_i) z(min_i) 1] ;
136                      if det(A0)<0
137                          min_i=i;
138                      end
139                  end
140              end
141          end
142      end
143      if min_i==0
144          edge=0;
145          break
```

```
146    end
147    A0=[0 0 0 1; x(i1) y(i1) z(i1) 1; x(i2) y(i2) z(i2) 1;
       x(min_i) y(min_i) z(min_i) 1];
148    V0=det(A0)/6;
149    if V0>0  % inner
150        NS=NS+1;
151        A=[0 0 0 1; x0(i1) y0(i1) z0(i1) 1; x0(i2) y0(i2)
       z0(i2) 1;  x0(min_i) y0(min_i) z0(min_i) 1];
152        V=V+abs(det(A))/24;
153        x11=x0(i1)/2; y11=y0(i1)/2; z11=z0(i1)/2;
154        x12=(x0(i1)+x0(i2))/3; y12=(y0(i1)+y0(i2))/3; z12=(z0(i1)+
       z0(i2))/3;
155        x13=(x0(i1)+x0(min_i)+x0(i2))/4; y13=(y0(i1)+y0(min_i)+
       y0(i2))/4; z13=(z0(i1)+z0(min_i)+z0(i2))/4;
156        x21=x0(i2)/2; y21=y0(i2)/2; z21=z0(i2)/2;
157        x22=(x0(i2)+x0(min_i))/3; y22=(y0(i2)+y0(min_i))/3;
       z22=(z0(i2)+z0(min_i))/3;
158        x23=(x0(i2)+x0(i1)+x0(min_i))/4; y23=(y0(i2)+y0(i1)+
       y0(min_i))/4; z23=(z0(i2)+z0(i1)+z0(min_i))/4;
159        x31=x0(min_i)/2; y31=y0(min_i)/2; z31=z0(min_i)/2;
160        x32=(x0(min_i)+x0(i1))/3; y32=(y0(min_i)+y0(i1))/3;
       z32=(z0(min_i)+z0(i1))/3;
161        x33=(x0(min_i)+x0(i2)+x0(i1))/4; y33=(y0(min_i)+y0(i2)+
       y0(i1))/4; z33=(z0(min_i)+z0(i2)+z0(i1))/4;
162        a13=sqrt((x11-x12)^2+(y11-y12)^2+(z11-z12)^2);
163        a11=sqrt((x12-x13)^2+(y13-y12)^2+(z13-z12)^2);
164        a12=sqrt((x11-x13)^2+(y11-y13)^2+(z11-z13)^2);
165        a23=sqrt((x21-x22)^2+(y21-y22)^2+(z21-z22)^2);
166        a21=sqrt((x22-x23)^2+(y23-y22)^2+(z23-z22)^2);
167        a22=sqrt((x21-x23)^2+(y21-y23)^2+(z21-z23)^2);
168        a33=sqrt((x31-x32)^2+(y31-y32)^2+(z31-z32)^2);
169        a31=sqrt((x32-x33)^2+(y33-y32)^2+(z33-z32)^2);
170        a32=sqrt((x31-x33)^2+(y31-y33)^2+(z31-z33)^2);
171        p3=a31+a32+a33; p3=p3/2; p1=(a11+a12+a13)/2; p2=(a21+a22+
       a23)/2;
172        area1=sqrt(p1*(p1-a11)*(p1-a12)*(p1-a13))*2;
173        area2=sqrt(p2*(p2-a21)*(p2-a22)*(p2-a23))*2;
174        area3=sqrt(p3*(p3-a31)*(p3-a32)*(p3-a33))*2;
175        l1(1)=(y12-y11)*(z13-z11)-(y13-y11)*(z12-z11);
176        l1(2)=(z12-z11)*(x13-x11)-(z13-z11)*(x12-x11);
177        l1(3)=(x12-x11)*(y13-y11)-(x13-x11)*(y12-y11);
178        l2(1)=(y22-y21)*(z23-z21)-(y23-y21)*(z22-z21);
179        l2(2)=(z22-z21)*(x23-x21)-(z23-z21)*(x22-x21);
180        l2(3)=(x22-x21)*(y23-y21)-(x23-x21)*(y22-y21);
181        l3(1)=(y32-y31)*(z33-z31)-(y33-y31)*(z32-z31);
```

```
182          l3(2)=(z32—z31)*(x33—x31)—(z33—z31)*(x32—x31);
183          l3(3)=(x32—x31)*(y33—y31)—(x33—x31)*(y32—y31);
184          ll1=sqrt(l1(1)^2+l1(2)^2+l1(3)^2);
185          ll2=sqrt(l2(1)^2+l2(2)^2+l2(3)^2);
186          ll3=sqrt(l3(1)^2+l3(2)^2+l3(3)^2);
187          l1=l1/ll1; l2=l2/ll2; l3=l3/ll3;
188          surf_area(i1,1:3)=surf_area(i1,1:3)—area1*l1;
189          surf_area(i2,1:3)=surf_area(i2,1:3)—area2*l2;
190          surf_area(min_i,1:3)=surf_area(min_i,1:3)—area3*l3;
191          surface0=sort([i1,i2,min_i]);
192          surface(NS, 1:3)=surface0;
193          if edge_used~=0
194              edge_used=[edge_used; edge(1,:)];
195          else
196              edge_used=edge(1,:);
197          end
198          edge(1,:)=[]; edge_particle(1)=[];
199          belong_i1=0; belong_i2=0;
200          for j=1:length(Edge(:,1)) % examine whether [i1 min] [min
     i2] in edge?
201              if (Edge(j,1)==i1&&Edge(j,2)==min_i)||(Edge(j,1)==
     min_i&&Edge(j,2)==i1)
202                  belong_i1=1;
203              end
204              if (Edge(j,1)==min_i&&Edge(j,2)==i2)||(Edge(j,2)==
     min_i&&Edge(j,1)==i2)
205                  belong_i2=1;
206              end
207          end
208          if belong_i1==0
209              edge=[edge; i1 min_i];  edge_particle=[edge_particle;
     i2];
210          else
211              for i=1:length(edge(:,1))
212                  if (edge(i,1)==i1&&edge(i,2)==min_i)||(edge(i,2)==
     i1&&edge(i,1)==min_i)
213                      if edge_used~=0
214                          edge_used=[edge_used; edge(i,:)];
215                      else
216                          edge_used=edge(i,:);
217                      end
218                      edge(i,:)=[]; edge_particle(i)=[];
219                      break
220                  end
221              end
```

```
222            end
223            if belong_i2==0
224                edge=[edge; min_i i2];    edge_particle=[edge_particle;
       i1];
225            else
226                for i=1:length(edge(:,1))
227                    if (edge(i,1)==i2&&edge(i,2)==min_i)||(edge(i,2)==
       i2&&edge(i,1)==min_i)
228                        if edge_used~=0
229                            edge_used=[edge_used; edge(i,:)];
230                        else
231                            edge_used=edge(i,:);
232                        end
233                        edge(i,:)=[]; edge_particle(i)=[];
234                        break
235                    end
236                end
237            end
238            Edge=[Edge; i1 min_i; min_i i2];
239        else
240            if edge_used~=0
241                edge_used=[edge_used; edge(1,:)];
242            else
243                edge_used=edge(1,:);
244            end
245            edge(1,:)=[];   edge_particle(1)=[];
246            for i=1:length(edge(:,1))
247                if(edge(i,1)==i1&&edge(i,2)==min_i)||(edge(i,2)==
       i1&&edge(i,1)==min_i)
248                    if edge_used~=0
249                        edge_used=[edge_used; edge(i,:)];
250                    else
251                        edge_used=edge(i,:);
252                    end
253                    edge(i,:)=[];    edge_particle(i)=[];
254                    break
255                end
256            end
257            for i=1:length(edge(:,1))
258                if (edge(i,1)==i2&&edge(i,2)==min_i)||(edge(i,2)==i2&&
       edge(i,1)==min_i)
259                    if edge_used~=0
260                        edge_used=[edge_used; edge(i,:)];
261                    else
262                        edge_used=edge(i,:);
```

```
263                     end
264                     edge(i,:)=[];   edge_particle(i)=[];
265                     break
266               end
267           end
268       end
269 end
270 x=x0; y=y0; z=z0;
271 E11=0; E12=0; E13=0; E21=0; E22=0; E23=0; E31=0; E32=0; E33=0;
272 for j=1:N
273     E11=E11+x(j)*surf_area(j,1); E12=E12+y(j)*surf_area(j,1); E13=
    E13+z(j)*surf_area(j,1);
274     E21=E21+x(j)*surf_area(j,2); E22=E22+y(j)*surf_area(j,2); E23=
    E23+z(j)*surf_area(j,2);
275     E31=E31+x(j)*surf_area(j,3); E32=E32+y(j)*surf_area(j,3); E33=
    E33+z(j)*surf_area(j,3);
276 end
277 CCz=(E32-E12*E31/E11)/(E22-E12*E21/E11);
    CCy=(E33-E13*E31/E11)/(E23-E13*E21/E11);
    CCx=(E13-E23*E12/E22)/(E33-E23*E32/E22);
278 Cz=E33-E13*E31/E11-(E23-E13*E21/E11)*CCz;
    Cy=E32-E12*E31/E11-(E22-E12*E21/E11)*CCy;
    Cx=E11-E21*E12/E22-(E31-E21*E32/E22)*CCx;
279 dsz=((surf_area(:,3)-surf_area(:,1)*E31/E11)-(surf_area(:,2)-
    surf_area(:,1)*E21/E11)*CCz)/Cz;
280 dsy=((surf_area(:,3)-surf_area(:,1)*E31/E11)-(surf_area(:,2)-
    surf_area(:,1)*E21/E11)*CCy)/Cy;
281 dsx=((surf_area(:,1)-surf_area(:,2)*E12/E22)-(surf_area(:,3)-
    surf_area(:,2)*E32/E22)*CCx)/Cx;
282 %%%%%%%%%%%%%%%%%%%%%% plot supporting particles
    %%%%%%%%%%%%%%%%%%%%%%
283 ox=[xt;x(1)]; oy=[yt;y(1)]; oz=[zt; z(1)];
284 plot3(x,y,z,'bo','MarkerFaceColor','b','markersize',10); hold on;
285 plot3(xt,yt,zt,'ro','MarkerFaceColor','r','markersize',10); hold
    on;
286 NN=length(surface(:,1));
287 set(gca,'FontName','times new Roman','FontSize',18);
288 axis equal; grid on;
289 xlabel('x'); ylabel('y'); zlabel('z'); legend('Supporting
    particles','target particle')
290 end
```

9.3 GSM simulation of Poiseuille flow

The GSM for simulating fluid dynamics problems using a globally unstructured grid is demonstrated by applying it to planar Poiseuille flow. The accuracy of the GSM solution is verified by comparing to the existing analytical solution from the theory of fluid mechanics:

$$U_{\text{final}} = \frac{P}{2\mu}\left(H^2 - y^2\right), \text{ where } \mu \text{ is dynamic viscosity} \tag{9.4}$$

As shown in Figure 9.6, the domain of the planar Poiseuille flow is defined as a planar tube of $L \times H$, with periodical boundaries on the left inlet and right outlet and zero-velocity boundary condition (no-slip BCs) on the top and bottom boundaries. The flow is driven by a constant body force P in x direction.

The governing equation of incompressible *laminar* flow under an Eulerian frame is derived from the continuity equation and the Navier–Stokes equation:

$$\frac{\partial u}{\partial t} = -\frac{1}{\rho}\frac{dp}{dx} + v\frac{\partial^2 u}{\partial y^2}, \text{ where } v = \frac{\mu}{\rho} \text{ is kinematic viscosity of fluid}$$

$$\tag{9.5}$$

The derivative terms in the governing equation are approximated with the GSM gradient operator using a global unstructured grid. The GSM solution and FDM solution obtained from the provided source code are plotted in Figures 9.7 and 9.8, respectively.

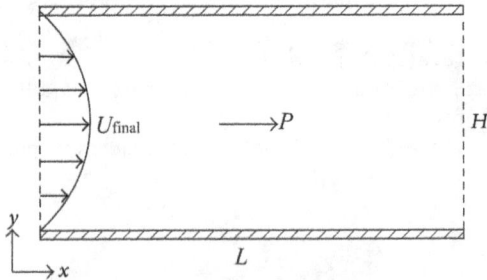

Figure 9.6. Geometry sketch of planar Poiseuille flow. $P = -dp/dx$ stands for the body force driving the flow in x direction, U_{final} is the final steady velocity.

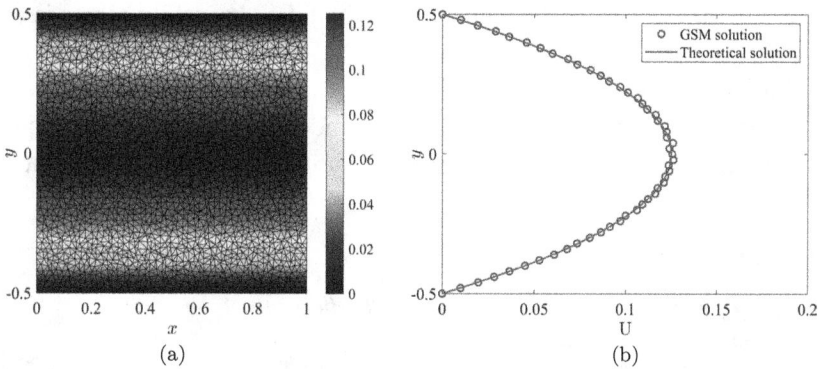

Figure 9.7. Velocity contour of flow by GSM (left) and comparison of GSM solution to theoretical solution (right).

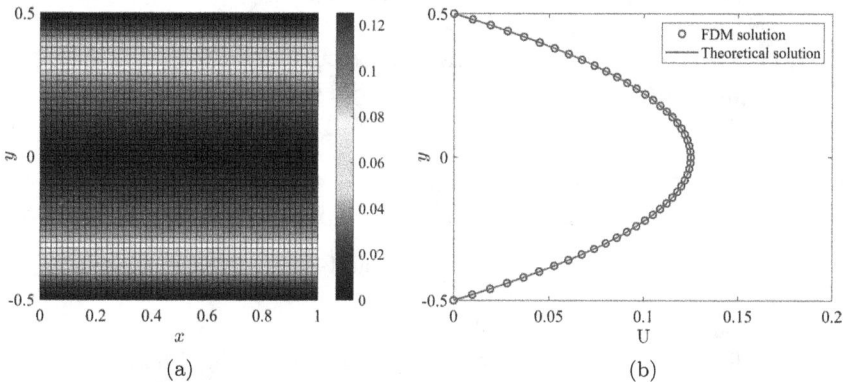

Figure 9.8. Velocity contour of flow by FDM (a) and comparison of FDM solution to theoretical solution (b).

Use of source code: The attached source code enables users to carry out the following explorations: (a) verify the effectiveness of the GSM framework when simulating dynamics flows under unstructured mesh by comparing to FDM; (b) study the influence of flow properties on the flow dynamics by using different values of parameters defined between lines 7 and 8 in the main program "GSM_Poiseuelle_flow.m"; (c) evaluate the accuracy and computational efficiency of GSM when employing a finer grid in simulation by changing the value of "N" in line 5. Note that, as the mesh resolution is changed, the critical time step Δt in line 9 is revised automatically to ensure the stability of the numerical method. So, the stability issue is not of concern.

Moreover, the source code may be extended to other incompressible flows, e.g., lid-driven cavity flow, by modifying the governing equation, computational domain, boundary conditions, and initial conditions properly.

Structure of source code

Outline of Main program GSM_Poiseuelle_flow.m
  Necessary comments and initialization % Line 1–3
    Define necessary controlling parameters % Line 4–11
    Define domain grid % Line 12–24
    Specify the initial condition and boundary conditions % Line 25–35
    Update velocity of flow based governing equation % Line 36–46
      Outline of GSM_gradient.m at line 42 and line 44
        Necessary comments and initialization % Line 2–7
        Calculate n*dS in the GSM gradient operator % Line 8–25
        Approximate gradient with standard GSM operator % Line 27
    Calculate difference of two subsequent velocity fields and print % Line 47–51
    Plot velocity contour % 52–58
    Compare GSM solution to theoretical solution % 59–64
Outline of Main program FDM_Poiseuelle_flow.m

  Necessary comments and initialization % Line 1–3
  Define necessary controlling parameters % Line 4–11
  Define domain grid % Line 12–18
  Specify the initial condition and boundary conditions % Line 19–22
  Update velocity of flow-based governing equation % Line 23–30
  Calculate difference of two subsequent velocity fields and print % Line 33–35
  Plot velocity contour % Line 36–42
  Compare FDM solution to theoretical solution % 43–48

The source code including the main program "GSM_Poiseuelle flow.m", function "GSM_gradient.m", and the main program "FDM_ Poiseuelle flow.m" is presented as follows.

Main program: GSM_Poiseuelle_flow.m

This program solves the incompressible Poiseuille flow with the Gradient Smoothing Method under the Eulerian frame. The gradient and Laplacian of field variables in the governing PDE are approximated with the standard GSM gradient operator. The velocity contour representing the dynamical process of flow is plotted out periodically. Also, the velocity from FDM solution along a selected vertical line is compared to the corresponding theoretical solution.

This program invokes the following function:

- **GSM_gradient.m**

Listing:

```
1   % This code is written to solve the 2D Poiseuelle flow by using
2   % Gradient Smoothing Method in Eulerian frame.
3   clear all; clc; clf;
4   %%%%%%%%% controlling parameters %%%%%%%%
5   N=101; % amount of nodes on one edge of the square domain
6   ds=1/(N-1); % grid size
7   P=1; % -dpdx pressure gradient term
8   mu=1; nu=mu/1; % % mu <=1 is dynamic viscosity of fluid; rho = 1
9   dt=ds^2*0.5; % time interval
10  writeInterval=1000; % time-step interval for plotting results
11  criterion_u=1.0e-7; % convergence criterion of velocity
    calculation
12  %%%%%%%%% define domain grid %%%%%%%%
13  X=[0:ds:1]; X=[X X]; Y=[zeros(1,N) ones(1,N)]; % generate top and
    bottom no-slip nodes
14  X=[X zeros(1,N-2) ones(1,N-2)]; % generate x coordinates of
    periodical boundary nodes
15  Y=[Y [ds:ds:1-ds] [ds:ds:1-ds]]; % generate y coordinates of
    periodical boundary nodes
16  x=zeros(N-2,N-2);   y=zeros(N-2,N-2); % irregularly distributed
    interior nodes
17  dx=0.4*(rand(N-2)-0.5)*2*ds; dy=0.4*(rand(N-2)-0.5)*2*ds; % noise
    assigned to uniform nodes
18  for i=2:N-1
19      for j=2:N-1
20          x(i-1,j-1)=(j-1)/(N-1); y(i-1,j-1)=(i-1)/(N-1); % uniform
    interior nodes
21      end
22  end
23  x=x+dx; y=y+dy; % generate irregularly distributed interior nodes
24  X=[X reshape(x,[1,(N-2)*(N-2)])]; Y=[Y reshape(y,[1,(N-2)*
    (N-2)])]; Y=Y-0.5; % reshape
25  %%%%%%%%%%%% Initial and boundary conditions %%%%%%%%%%%
26  u=zeros(length(X),1); % initial condition of velocity u
27  index_L=[1 2*N+1:3*N-2 N+1]; % index array of left boundary nodes
28  index_R=[N 3*N-1:4*N-4 2*N]; % index array of right boundary nodes
```

```
29  index_L_virt=[N*N+1:N*N+N]; % index array of left virtual nodes
30  index_R_virt=[N*N+N+1:N*N+2*N]; % index array of right virtual
    nodes
31  X=[X X(index_L)+1.0+ds 1.0-X(index_R)-ds]; % assign virtual
    boundary nodes
32  Y=[Y Y(index_L) Y(index_R)]; % assign virtual boundary nodes
33  T=delaunay(X,Y); %%%%%%%% generate unstructured mesh via Delaunay
    algorithm
34  u(1:2*N)=0.0; % no-slip BCs for top and bottom boundary
35  u(index_L_virt)=u(index_L); u(index_R_virt)=u(index_R); % Periodic
    left and right BCs
36  %%%%%%%%%%%% velocity update %%%%%%%%%%
37  time=0; s=0;  t=0;  iter=0;     Res_u=1.0;
38  while (Res_u>criterion_u)
39      uold=u;   iter=iter+1;   t=t+dt;
40      [u_x,u_y]=GSM_gradient(T,X,Y,length(X),u); % calculate
    velocity gradient
41      u_y(index_L_virt)=u_y(index_L); u_y(index_R_virt)=u_y
    (index_R); % Periodic BCs
42      [u_xy,u_yy]=GSM_gradient(T,X,Y,length(X),u_y); % calculate
    velocity u_yy
43      u_yy(index_L_virt)=u_yy(index_L); u_yy(index_R_virt)=u_yy
    (index_R); % Periodic BCs
44      u(2*N+1:N*N)=u(2*N+1:N*N)+dt*(P+u_yy(2*N+1:N*N)*nu); % update
    u of interior domain
45      u(1:2*N)=0.0; % no-slip BCs for top and bottom boundary
46      u(index_L_virt)=u(index_L); u(index_R_virt)=u(index_R);
    % Periodic left and right BCs
47      if iter/writeInterval == round(iter/writeInterval)
48          s=s+1;  times(s)=t;
49          res_u(s)=max(max(abs(u-uold))); % maximum residual of
    increament of velocity
50          Res_u=res_u(s);
51          fprintf('Calculating ... time = %0.2f | Res_u-7=%0.2f \n',
    t, log10(Res_u))
52          hh=suptitle(sprintf('Res_u-7=%0.2f | Time =%0.3f sec',
    log10(Res_u),t));
53          subplot(1,2,1) %%%%%%%% plot velocity contour
54          h=trimesh(T,X,Y,u,'FaceColor','interp','EdgeColor','k');
55          view(2); axis equal; axis([0 1 -0.5 0.5])
56          title('U'); xlabel('x');  ylabel('y')
57          colorbar;  colormap jet; caxis([0 P/2/mu*0.5^2]);
58          set(gca, 'Fontname', 'Times New Roman','FontSize',15);
59          subplot(1,2,2) %%%%%%%% compare numerical solution to
    exact solution
```

```
60          plot(u(index_L),Y(index_L),'o',P/2/mu*(0.5^2−
    Y(index_L).^2),Y(index_L),'−','linewidth',1.5);
61          legend('GSM solution','Exact solution'); xlabel('U');
    ylabel('y');
62          drawnow;
63      end
64  end
```

GSM_gradient.m

This function approximates the gradient with the standard GSM gradient operator with the help of a global mesh represented by a T matrix.

Listing:

```
1   function [dFx,dFy] = GSM_gradient (T, x, y, N, F)
2   % This function is written to approximate gradient of scalar
    function
3   % F by GSM gradient operator based on n−GSD.
4   %%%%% initialize necessary matrices %%%%%%
5   NN=length(T); % total number of triangles representing the whole
    domain
6   dSx_F=zeros(N,1); dSy_F=zeros(N,1); % DeltaS*U in the GSM gradient
    operator
7   V=zeros(N,1); % volume or area of GSD in the GSM gradient operator
8   %%%%% calculate n*dS of GSM gradient operator %%%%%%%%%%
9   for t=1:NN
10      i1=T(t,1); i2=T(t,2); i3=T(t,3); % indices are ordered in
    anticlockwise sequence
11      area=abs((x(i2)−x(i1))*(y(i3)−y(i1)) − (x(i3)−x(i1))*(y(i2)−y(
    i1)))/6;
12      V(i1)=V(i1)+area; V(i2)=V(i2)+area; V(i3)=V(i3)+area; % area
    of GSD
13      dSx_F(i1)=dSx_F(i1)+(y(i3)/3.0−y(i1)/6.0−y(i2)/6.0)*(F(i2)+F(
    i1))/2.0;
14      dSy_F(i1)=dSy_F(i1)+(x(i3)/3.0−x(i1)/6.0−x(i2)/6.0)*(F(i2)+F(
    i1))/2.0;
15      dSx_F(i1)=dSx_F(i1)+(y(i1)/6.0+y(i3)/6.0−y(i2)/3.0)*(F(i3)+F(
    i1))/2.0;
16      dSy_F(i1)=dSy_F(i1)+(x(i1)/6.0+x(i3)/6.0−x(i2)/3.0)*(F(i3)+F(
    i1))/2.0;
17      dSx_F(i2)=dSx_F(i2)+(y(i1)/3.0−y(i2)/6.0−y(i3)/6.0)*(F(i3)+F(
    i2))/2.0;
18      dSy_F(i2)=dSy_F(i2)+(x(i1)/3.0−x(i2)/6.0−x(i3)/6.0)*(F(i3)+F(
    i2))/2.0;
19      dSx_F(i2)=dSx_F(i2)+(y(i1)/6.0+y(i2)/6.0−y(i3)/3.0)*(F(i1)+F(
    i2))/2.0;
```

```
20      dSy_F(i2)=dSy_F(i2)+(x(i1)/6.0+x(i2)/6.0-x(i3)/3.0)*(F(i1)+F(
        i2))/2.0;
21      dSx_F(i3)=dSx_F(i3)+(y(i2)/3.0-y(i1)/6.0-y(i3)/6.0)*(F(i1)+F(
        i3))/2.0;
22      dSy_F(i3)=dSy_F(i3)+(x(i2)/3.0-x(i1)/6.0-x(i3)/6.0)*(F(i1)+F(
        i3))/2.0;
23      dSx_F(i3)=dSx_F(i3)+(y(i3)/6.0+y(i2)/6.0-y(i1)/3.0)*(F(i2)+F(
        i3))/2.0;
24      dSy_F(i3)=dSy_F(i3)+(x(i3)/6.0+x(i2)/6.0-x(i1)/3.0)*(F(i2)+F(
        i3))/2.0;
25  end
26  %%%% approximate gradient with GSM gradient operator %%%%%
27  dFx=dSx_F./V; dFy=-dSy_F./V; % GSM gradient operator for gradient
    components of F
28  end
```

Main program: FDM_Poiseuelle_flow.m

This program solves the incompressible Poiseuille flow with the Finite Difference Method under the Eulerian frame. The gradient and Laplacian of field variables in the governing PDE are approximated with the 2nd-order accurate central-difference scheme. The velocity contour representing the dynamical process of flow is plotted out periodically. Also, the velocity from the FDM solution along a selected vertical line is compared to the corresponding theoretical solution.

Listing:

```
1   % This code is written to solve the 2D Poiseuelle flow by using
2   % the Finite Difference Scheme.
3   clear all; clc; clf;
4   %%%%%%%%%% controlling parameters %%%%%%%
5   N=101; % amount of nodes on one edge of the square domain
6   ds=1/(N-1); % grid size
7   P=1; % -dpdx pressure gradient term
8   mu=1; nu = mu/1; % mu <=1 is dynamic viscosity of fluid; rho = 1
9   dt=ds^2*0.5; % time interval
10  writeInterval=1000; % time-step interval for plotting results
11  criterion_u=1.0e-7; % convergence criterion of velocity
    calculation
12  %%%%%%%%%% define domain grid %%%%%%%%%
13  x=zeros(N,N);   y=zeros(N,N);
14  for i=1:N
```

```
15      for j=1:N
16          x(i,j)=(j-1)/(N-1);  y(i,j)=(i-1)/(N-1)-0.5;
17      end
18  end
19  %%%%%%%%%%%%%% ICs and BCs  %%%%%%%%%%
20  u=zeros(N,N); % initial condition settings
21  u(1,:)=0.0; u(N,:)=0.0; % no-slip boundary condition
22  time=0; s=0; t=0; iter=0;    Res_u=1.0;
23  %%%%%%%%%%%%%% velocity update %%%%%%%%%%
24  while (Res_u>criterion_u)
25      uold=u;   iter=iter+1; t=t+dt;
26      u_yy=zeros(N,N); % second derivative of u w.r.t. y
27      u_n=u(1:N-2,:); u_s=u(3:N,:); % the upper and lower grid
28      u_yy=(u_n+u_s-2.0*u(2:N-1,:))/ds/ds; % compute u_xx with
central FD scheme
29      u(2:N-1,:)=u(2:N-1,:)+dt*(P+u_yy*nu); % update velocity based
governing equation
30      u(1,:)=0.0; u(N,:)=0.0; % no-slip boundary condition on top
and bottom
31      if iter/writeInterval == round(iter/writeInterval)
32          s=s+1;  times(s)=t;
33          res_u(s)=max(max(abs(u-uold))); % maximum residual of
increment of velocity
34          Res_u=res_u(s);
35          fprintf('Calculating ... time = %0.2f | Res_u-7=%0.2f \n',
t, log10(Res_u))
36          hh=suptitle(sprintf('Res_u-7=%0.2f | Time =%0.3f sec',
log10(Res_u),t));
37          subplot(1,2,1) %%%%%%% plot velocity contour
38          h=surf(x,y,u, 'facecolor', 'interp');
39          view(2); axis equal; axis([0 1 -0.5 0.5])
40          title('U'); xlabel('x');  ylabel('y')
41          colorbar;  colormap jet; caxis([0 P/2/mu*0.5^2]);
42          set(gca, 'Fontname', 'Times New Roman','FontSize',15);
43          subplot(1,2,2)  %%%%%%%% compare numerical solution to
exact solution
44          plot(u(:,1),y(:,1),'o',P/2/mu*(0.5^2-y(:,1).^2),y(:,1),'-'
,'linewidth',1.5);
45          legend('FDM solution','Exact solution'); xlabel('U');
ylabel('y');
46          drawnow;
47      end
48  end
```

9.4   L-GSM simulation of Couette flow

This section presents a benchmarking example, i.e., incompressible Couette flow, solved by L-GSM under the Lagrangian frame. The details about the numerical model and properties of flow can be found in Section 7.2.1 in Chapter 7, and the governing PDEs can be found in Section 7.1.

Use of source code: The attached source code may be used to carry out the following explorations: (a) verify the effectiveness of the proposed L-GSM (or Local-GSM more accurately) framework to simulate large deformation dynamics flows under the Lagrangian frame. To evaluate the accuracy of the L-GSM solution, the velocity along a vertical line from L-GSM is compared to the theoretical solution, as presented in Figure 9.9; (b) study the influence of flow properties on the flow dynamics by using different values of parameters defined between lines 9 and 12 in the main program "L-GSM_Couette_flow.m"; (c) evaluate the accuracy condition and computational efficiency of L-GSM, when employing more particles in simulation by changing the value of "Nx" in line 7. Typically, the numerical results from the provided Matlab source code should be consistent or

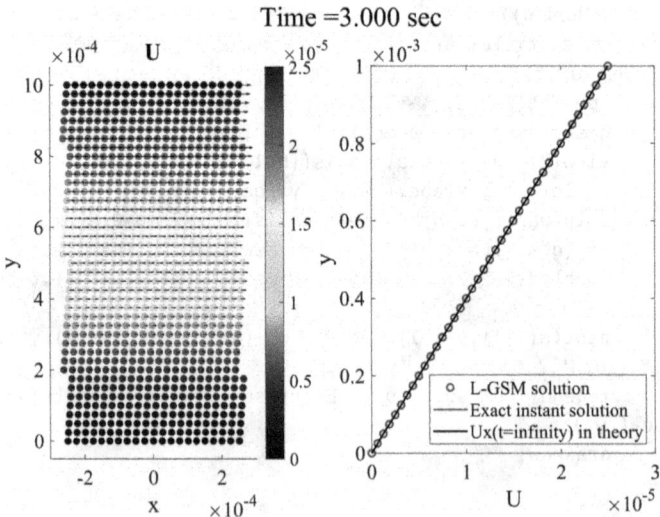

Figure 9.9.   Velocity vector of flow from L-GSM (left) and comparison of L-GSM solution to theoretical solution (right).

very close to those presented in Section 7.2.1, although those results are produced from a Fortran version of the code. Note that, as the number of particles is changed, the critical time step Δt in line 14 should also be corrected according to equation (6.59) in order to ensure the stability of the numerical method.

The source code is also easy to extend to other incompressible flows, e.g., Poiseuille flow, by slightly modifying the computational domain, boundary conditions, and initial conditions properly.

Structure of source code

Outline of L-GSM_Couette_flow.m
    Necessary comments and initialization % Line 1–4
    Define necessary controlling parameters % Line 5–15
    Generate initial configuration of particles % Line 16–25
    Apply the initial conditions % 26–29
    Update field variables and locations of particles % Line 30–64
        Outline of 'rho_particle.m' at line 46
        Find out supporting particles locally % Line 4–7
        Ordering supporting particles % Line 8–21
        Calculate area of n-GSD % Line 22–27
        Approximate gradient with GSM gradient operator % Line 28–43
        Calculate change rate of density based on continuity equation % Line 45
        Outline of 'velocity_particle.m' at line 47
        Find out supporting particles locally % Line 4–7
        Ordering supporting particles % Line 8–21
        Calculate area of n-GSD % Line 22–27
        Approximate gradient with GSM gradient operator % Line 28–43
        Calculate change rate of velocity from momentum equation % Line 45
        Outline of 'art_visc.m' at line 48
        Find out supporting particles locally % Line 4–7
        Ordering supporting particles % Line 8–21
        Calculate area of n-GSD % Line 22–28
        Approximate gradient with GSM gradient operator % Line 29–43
        Calculate acceleration caused by artificial viscosity % Line 44–45
    Plot velocity vector and locations of particles % Line 65–74
    Compare to velocity from L-GSM to theoretical solution % Line 75–88

The source code including the main program "L-GSM_Couette_flow.m" and functions "rho_partial.m", "velocity_partial.m", and "art_visc.m" are listed as follow.

Main program: L-GSM_Couette_flow.m

This program simulates the incompressible Couette flow with L-GSM under the Lagrangian frame. The gradient of field variables in the governing PDEs is approximated with a GSM gradient operator in a conservative fashion. The velocity vector and movement of particles are plotted out periodically. Meanwhile, the velocity from L-GSM solution along a selected vertical line is compared with the corresponding theoretical solution.

This program makes the call of the following functions:

- **rho_partial.m**
- **velocity_partial.m**
- **art_visc.m**

Listing:

```
1    % This code is written to solve the 2D Couette flow
2    % using Lagrangian Gradient Smoothing Method. For detailed
     information,
3    % please refer to "Zirui Mao, G.R. Liu, Int J Numer Methods Eng.
     2018;113:858—C890.".
4    clear all; clc; clf;
5    %%%%%%%%% controlling parameters %%%%%%%%
6    Lx=0.5e—3; Ly=1e—3; % computational domain
7    Nx=21; Ny=2*(Nx—1)+1; % amount of nodes on x, y edge of the domain
8    ds=Lx/(Nx—1); % grid size
9    rho0=1e3; % initial density
10   mu=1e—6; % kinetic viscosity
11   U0 = 2.5e—5; % Horizontal velocity of top lid
12   c = 1e—4; % sound speed in fluid
13   TimeStep=20000; % total time steps to run
14   dt=1e—4; % critical time step
15   writeInterval=200; % time—step interval for results plot
16   %%%%%%%%% generate initial particles %%%%%%%%
17   X=[0:ds:Lx]; X=[X X]; Y=[zeros(1,Nx) Ly*ones(1,Nx)]; % generate
        no—slip boundary particles
18   Nb=length(X); % amount of no—slip boundary particles
19   x=zeros(Ny—2,Nx);   y=zeros(Ny—2,Nx); % interior nodes
20   for i=1:Ny—2
21       for j=1:Nx
22           x(i,j)=(j—1)*ds; y(i,j)=(i)*ds; % initially uniform
     interior nodes
23       end
```

```
24  end
25  X=[X reshape(x,[1,(Ny-2)*(Nx)])]; X=X-Lx/2; Y=[Y reshape(y,[1,(Ny
    -2)*(Nx)])]; % reshape
26  %%%%%%%%%%%%%%% ICs %%%%%%%%%%
27  N=length(X); vx=zeros(1,N); vx(Nb/2+1:Nb)=U0; vy=zeros(1,N); rho=
    rho0*ones(1,N); pres=c*c*rho;
28  drho=zeros(1,N); dvelocity_x=zeros(1,N); dvelocity_y=zeros(1,N);
29  rho_min=rho; vx_min=vx; vy_min=vy;  % store the mid-step variable
    by using leaf-frog time scheme
30  %%%%%%%%%%%%%% update of field variable and locations %%%%%%%%%%%
31  time=0; s=0;  t=0;
32  for itimestep=1:TimeStep
33      t=t+dt;
34      if itimestep~=1
35          rho_min=rho; vx_min=vx; vy_min=vy;
36          rho=rho+dt/2*drho; pres=c*c*rho; vx=vx+dt/2*dvelocity_x;
    vy=vy+dt/2*dvelocity_y;
37          vx(1:Nb/2)=0.0; vy(1:Nb/2)=0.0; vx(Nb/2+1:Nb)=U0; vy(Nb
    /2+1:Nb)=0;
38      end
39      XX=[X X(1,1:2*Nx)]; YY=[Y -ds*ones(1,Nx) (Ly+ds)*ones(1,Nx)];
    % add a layer of virtual particles for no-slip BCs
40      vvx=[vx vx(1,1:2*Nx)]; vvy=[vy vy(1,1:2*Nx)]; rrho=[rho rho0*
    ones(1,Nx*2)];
41      index_virt_L=find(XX>Lx/2-1.5*ds); index_virt_R=find(XX<1.5*ds
    -Lx/2); % find particles applying Periodical BCs
42      XX=[XX XX(index_virt_L)-Lx-ds XX(index_virt_R)+Lx+ds];
    % Periodical BC assignment
43      YY=[YY YY(index_virt_L) YY(index_virt_R)];
44      vvx=[vvx vvx(index_virt_L) vvx(index_virt_R)];  vvy=[vvy vvy(
    index_virt_L) vvy(index_virt_R)];
45      rrho=[rrho rrho(index_virt_L) rrho(index_virt_R)];    ppres=c*c
    *rrho; % equation of state
46      [drrho]=rho_partial (XX,YY,length(XX),ds, vvx, vvy, rrho);
    % calculate partial rho in continuity equation
47      [dvelocity_px,dvelocity_py]=velocity_partial (XX,YY,length(XX)
    ,ds,ppres,rrho); % in momentum equation
48      [avx, avy] = art_visc (XX, YY, length(XX), ds, vvx, vvy, rrho,
    mu); % calculate viscosity-acceleration term
49      drho=drrho(1:N);% = right hand side of continuity equation
50      dvelocity_x=-(dvelocity_px(1:N)+avx(1:N)); % = right hand side
    of momentum equation in x
51      dvelocity_y=-(dvelocity_py(1:N)+avy(1:N)); % = right hand side
    of momentum equation in x
52      if itimestep==1 % update field variables and positions
```

```
53          rho=rho+dt/2*drho; pres=c*c*rho; vx=vx+dt/2*dvelocity_x;
     vy=vy+dt/2*dvelocity_y;
54          vx(1:Nb/2)=0.0; vy(1:Nb/2)=0.0; vx(Nb/2+1:Nb)=U0; vy(Nb
     /2+1:Nb)=0;
55          X=X+dt*vx; Y=Y+dt*vy;
56          X(X>Lx/2+ds/2)=X(X>Lx/2+ds/2)-Lx-ds; % periodical BC
     treatment
57          X(X<-Lx/2-ds/2)=X(X<-Lx/2-ds/2)+Lx+ds; % periodical BC
     treatment
58      else % update field variables and positions of particles
59          rho=rho_min+dt*drho; pres=c*c*rho; vx=vx_min+dt*
     dvelocity_x; vy=vy_min+dt*dvelocity_y;
60          vx(1:Nb/2)=0.0; vy(1:Nb/2)=0.0; vx(Nb/2+1:Nb)=U0; vy(Nb
     /2+1:Nb)=0;
61          X=X+dt*vx; Y=Y+dt*vy;
62          X(X>Lx/2+ds/2)=X(X>Lx/2+ds/2)-Lx-ds; % periodical BC
     treatment
63          X(X<-Lx/2-ds/2)=X(X<-Lx/2-ds/2)+Lx+ds; % periodical BC
     treatment
64      end
65      if itimestep/writeInterval == round(itimestep/writeInterval) %
     plot results
66          s=s+1; times(s)=t;
67          fprintf('Calculating ... time = %0.2f | percentage = %0.2f
     \n', t, itimestep/TimeStep*100.0)
68          subplot(1,2,1) %%%%%%%% plot velocity contour
69          scatter(X,Y,[],vx,'filled'); hold on;
70          quiver(X,Y,vx,vy); hold off;
71          view(2); axis equal; colorbar; colormap jet;
72          axis([-3.0e-4 3.0e-4 -0.5e-4 1.05e-3]);
73          title('U'); xlabel('x'); ylabel('y')
74          set(gca, 'Fontname', 'Times New Roman','FontSize',18);
75          subplot(1,2,2) %%%%%%%%% compare numerical solution to
     exact solution
76          vx_plot=[vx(1) vx(Nb+1:Nb+Ny-2) vx(Nx+1)];
77          y_plot=[Y(1) Y(Nb+1:Nb+Ny-2) Y(Nx+1)];
78          vx_theory_infinity=U0/Ly*y_plot; vx_theory=
     vx_theory_infinity;
79          for n=1:10
80              vx_theory=vx_theory+2*U0/n/pi*(-1)^n*sin(n*pi*y_plot/
     Ly)*exp(-mu*n^2*pi^2*t/Ly^2);
81          end
82          plot(vx_plot,y_plot,'o',vx_theory,y_plot,'-',
     vx_theory_infinity,y_plot,'k','linewidth',1.5);
```

```
83          legend('L–GSM solution','Exact instant solution', 'Ux(t=
    infinity) in theory','location','southeast'); xlabel('U'); ylabel
    ('y');
84          set(gca, 'Fontname', 'Times New Roman','FontSize',18);
85          drawnow;
86          hh=suptitle(sprintf('Time =%0.3f sec',t));
87      end
88  end
```

rho_partial.m

This function calculates the change rate of density from a continuity equation by using the GSM gradient operator in a consistent and conservative manner.

Listing:

```
1   function [drho] = rho_partial (X, Y, NN, ds, vx, vy, rho)
2   dfx=zeros(1,NN); dfy=zeros(1,NN);
3   for k=1:NN % for all real particles
4       %%%%%% find out the neighboring particles %%%%%%%%%%%%
5       distance=sqrt((X–X(k)).^2+(Y–Y(k)).^2); % distance from other
    particles
6       L=find(distance<=1.5*ds); % supporting particles list
7       L(L==k)=[];
8       %%%%%% ordering supporting particles anti–clockwise %%%%%
    %%%
9       N=length(L);
10      P_local=[X(L)–X(k); Y(L)–Y(k)];
11      angle=zeros(1,N);
12      for i=1:N
13          if P_local(2,i)>=0
14              angle(i)=acosd(P_local(1,i)/distance(L(i)));
15          else
16              angle(i)=360–acosd(P_local(1,i)/distance(L(i)));
17          end
18      end
19      [angle,I]=sort(angle);
20      L=L(I);
21      x=X(L); y=Y(L);
22      %%%%% calculate area of n–GSD %%%%%%%%%
23      area=(x(N)*y(1)–x(1)*y(N))/6;
24      o=[X(k); Y(k)];
25      for j=1:N–1
26          area=area+(x(j).*y(j+1)–x(j+1).*y(j))/6; % calculate the
    area V of n–GSD
```

```
27     end
28     %%%%%%%%%%%%% gradient approximation %%%%%%%%%%%%%%%
29     for j=1:N
30         jm1=j-1;jp1=j+1;
31         if j==1
32             jm1=N;
33         elseif j==N
34             jp1=1;
35         end
36         x_up=(x(jp1)+x(j)+o(1))/3; y_up=(y(jp1)+y(j)+o(2))/3;
37         dSx=y_up-(o(2)+y(j))/2; dSy=x_up-(o(1)+x(j))/2;
38         x_low=(x(jm1)+x(j)+o(1))/3; y_low=(y(jm1)+y(j)+o(2))/3;
39         dSx=dSx-y_low+(o(2)+y(j))/2; dSy=dSy-x_low+(o(1)+x(j))/2;
40         dfx(k)=dfx(k)+(vx(L(j))-vx(k))/2.*dSx/area;
41         dfy(k)=dfy(k)-(vy(L(j))-vy(k))/2.*dSy/area;
42     end
43  end
44  %%%% calculate Drho/Dt in the continuity equation %%%%%%%%%%%
45  drho=(-dfx-dfy).*rho; % right hand term of the continuity equation
46  end
```

Function: velocity_partial

This function calculates the pressure-caused change rate of velocity from a momentum equation by using the GSM gradient operator in a consistent and conservative manner.

Listing:

```
1   function [dvx,dvy] = velocity_partial (X, Y, NN, ds, P, rho)
2   dfx=zeros(1,NN); dfy=zeros(1,NN);
3   for k=1:NN % for all real particles
4       %%%%%% find out the neighboring particles %%%%%%%%%%%
5       distance=sqrt((X-X(k)).^2+(Y-Y(k)).^2); % distance from other
    particles
6       L=find(distance<=1.5*ds); % supporting particles list
7       L(L==k)=[];
8       %%%%%% ordering supporting particles anti-clockwise %%%%%%%%
9       N=length(L);
10      P_local=[X(L)-X(k); Y(L)-Y(k)];
11      angle=zeros(1,N);
12      for i=1:N
13          if P_local(2,i)>=0
14              angle(i)=acosd(P_local(1,i)/distance(L(i)));
15          else
16              angle(i)=360-acosd(P_local(1,i)/distance(L(i)));
```

```
17              end
18          end
19          [angle,I]=sort(angle);
20          L=L(I);
21          x=X(L); y=Y(L);
22          %%%%% calculate area of n-GSD %%%%%%%%%
23          area=(x(N).*y(1)-x(1).*y(N))/6;
24          o=[X(k); Y(k)];
25          for j=1:N-1
26              area=area+(x(j).*y(j+1)-x(j+1).*y(j))/6; % calculate the
   area V of n-GSD
27          end
28          %%%%%%%%%%%%%% gradient approximation %%%%%%%%%%%%%%%%%%
29          for j=1:N
30              jm1=j-1;jp1=j+1;
31              if j==1
32                  jm1=N;
33              elseif j==N
34                  jp1=1;
35              end
36              x_up=(x(jp1)+x(j)+o(1))/3; y_up=(y(jp1)+y(j)+o(2))/3;
37              dSx=y_up-(o(2)+y(j))/2; dSy=x_up-(o(1)+x(j))/2;
38              x_low=(x(jm1)+x(j)+o(1))/3; y_low=(y(jm1)+y(j)+o(2))/3;
39              dSx=dSx-y_low+(o(2)+y(j))/2; dSy=dSy-x_low+(o(1)+x(j))/2;
40              dfx(k)=dfx(k)+(P(L(j))+P(k))/2.*dSx/area;
41              dfy(k)=dfy(k)-(P(L(j))+P(k))/2.*dSy/area;
42          end
43      end
44      %%%%%%%%% calculate pressure caused acceleration %%%%%%%%%%
45      dvx=-dfx./rho; dvy=-dfy./rho; % the first term in the right hand
   side of momentum equation
46  end
```

art_visc.m

This function calculates the viscosity-caused change rate of velocity from the momentum equation by using the GSM gradient operator in a consistent and conservative manner.

Listing:

```
1   function [avx,avy] = art_visc (X, Y, NN, ds, vx, vy, rho, mu)
2   avx=zeros(1,NN); avy=zeros(1,NN);
3   for k=1:NN % for all real particles
4       %%%%%% find out the neighboring particles %%%%%%%%%%
```

```
5       distance=sqrt((X-X(k)).^2+(Y-Y(k)).^2); % distance from other
    particles
6       L=find(distance<=1.5*ds); % supporting particles list
7       L(L==k)=[];
8       %%%%%% ordering supporting particles anti-clockwise %%%%%%%%
9       N=length(L);
10      P_local=[X(L)-X(k); Y(L)-Y(k)];
11      angle=zeros(1,N);
12      for i=1:N
13          if P_local(2,i)>=0
14              angle(i)=acosd(P_local(1,i)/distance(L(i)));
15          else
16              angle(i)=360-acos(P_local(1,i)/distance(L(i)));
17          end
18      end
19      [angle,I]=sort(angle);
20      L=L(I);
21      x=X(L); y=Y(L);
22      %%%%% calculate area of n-GSD %%%%%%%%%
23      area=(x(N).*y(1)-x(1).*y(N))/6;
24      o=[X(k); Y(k)];
25      DIST=(o(1)-x).^2+(o(2)-y).^2;
26      for j=1:N-1
27          area=area+(x(j).*y(j+1)-x(j+1).*y(j))/6; % calculate the
    area V of n-GSD
28      end
29      %%%%%%%%%%%%%% gradient approximation %%%%%%%%%%%%%%%%
30      for j=1:N
31          jm1=j-1;jp1=j+1;
32          if j==1
33              jm1=N;
34          elseif j==N
35              jp1=1;
36          end
37          x_up=(x(jp1)+x(j)+o(1))/3; y_up=(y(jp1)+y(j)+o(2))/3;
38          dSx=y_up-(o(2)+y(j))/2; dSy=x_up-(o(1)+x(j))/2;
39          x_low=(x(jm1)+x(j)+o(1))/3; y_low=(y(jm1)+y(j)+o(2))/3;
40          dSx=dSx-y_low+(o(2)+y(j))/2; dSy=dSy-x_low+(o(1)+x(j))/2;
41          dfx=(o(1)-x(j))*dSx/area;
42          dfy=-(o(2)-y(j))*dSy/area;
43          %%%%%%%% calculate the viscosity-caused acceleration
    %%%%%%%%%%%
44          avx(k)=avx(k)-mu*(rho(k)+rho(L(j)))/2/rho(k)/DIST(j)*(dfx+
    dfy)*(vx(k)-vx(L(j)));
```

```
45        avy(k)=avy(k)—mu*(rho(k)+rho(L(j)))/2/rho(k)/DIST(j)*(dfx+
   dfy)*(vy(k)—vy(L(j)));
46      end
47   end
48   end
```

Index

9 789811 280009